Studies in Modern Chemistry

Advanced courses in chemistry are changing rapidly in both structure and content. The changes have led to a demand for up-to-date books that present recent developments clearly and concisely. This series is meant to provide advanced students with books that will bridge the gap between the standard textbook and research paper. The books should also be useful to a chemist who requires a survey of current work outside his own field of research. Mathematical treatment has been kept as simple as is consistent with clear understanding of the subject.

Careful selection of authors actively engaged in research in each field, together with the guidance of four experienced editors, has ensured that each book ideally suits the needs of persons seeking a comprehensible and modern treatment of rapidly developing areas of chemistry.

William C. Agosta, The Rockefeller University

R. S. Nyholm, FRS, University College London

Consulting Editors

Academic editor for this volume

Allan Maccoll, University College London

Studies in Modern Chemistry

R. L. M. Allen
Colour Chemistry

R. B. Cundall and A. Gilbert
Photochemistry

T. L. Gilchrist and C. W. Rees
Carbenes, Nitrenes, and Arynes

S. F. A. Kettle
Coordination Compounds

Ruth M. Lynden-Bell and Robin K. Harris
Nuclear Magnetic Resonance Spectroscopy

A. G. Maddock
Lanthanides and Actinides

M. F. R. Mulcahy
Gas Kinetics

E. S. Swinbourne
Analysis of Kinetic Data

T. C. Waddington
Non-aqueous Solvents

K. Wade
Electron Deficient Compounds

Analysis of Kinetic Data

E. S. Swinbourne
New South Wales Institute of Technology

Nelson

Thomas Nelson and Sons Ltd
36 Park Street London W1Y 4DE
PO Box 18123 Nairobi Kenya

Thomas Nelson (Australia) Ltd
597 Little Collins Street Melbourne 3000

Thomas Nelson and Sons (Canada) Ltd
81 Curlew Drive Don Mills Ontario

Thomas Nelson (Nigeria) Ltd
PO Box 336 Apapa Lagos

Thomas Nelson and Sons (South Africa) (Proprietary) Ltd
51 Commissioner Street Johannesburg

First published in Great Britain 1971
ISBN 978-1-4684-7687-3   ISBN 978-1-4684-7685-9 (eBook)
DOI 10.1007/978-1-4684-7685-9

Illustrations by Colin Rattray & Associates
Filmset by Keyspools Ltd Golborne Lancs

Contents

Preface

Data analysis is important from two points of view: first, it enables a large mass of information to be reduced to a reasonable compass, and second, it assists in the interpretation of experimental results against some framework of theory. The purpose of this text is to provide a practical introduction to numerical methods of data analysis which have application in the field of experimental chemical kinetics.

Recognizing that kinetic data have many features in common with data derived from other sources, I have considered it appropriate to discuss a selection of general methods of data analysis in the early chapters of the text. It is the author's experience that an outline of these methods is not always easy to locate in summary form, and that their usefulness is often not sufficiently appreciated. Inclusion of these methods in the early chapters has been aimed at simplifying discussion in the later chapters which are more particularly concerned with kinetic systems. By the provision of a number of worked examples and problems, it is hoped that the reader will develop a feeling for the range of methods available and for their relative merits.

Throughout the text, the mathematical treatment has been kept relatively simple, lengthy proofs being avoided. I have preferred to indicate the 'sense' and usefulness of the various methods rather than to justify them on strict mathematical grounds.

It has been assumed that the reader has some prior knowledge of chemical kinetics, and the book is therefore appropriate for use by undergraduate students of chemistry or chemical engineering, more particularly in their later course years, and by graduate students entering the research field of experimental kinetics.

In a book of this size, the choice of topics and the extent of discussion are necessarily restricted; a list of references for further reading has been included with each chapter, on the understanding that some readers will wish to extend their knowledge considerably in certain areas, while others will seek a broad but more limited extension of their interest.

ELLICE S. SWINBOURNE

1 Repeated observations

Knowledge in science is built upon observations. These observations may be largely qualitative in nature at the original or exploratory stage of a study, but attempts are usually made to represent them in the form of quantitative data at some later stage. It is important that experimental information should be collected and catalogued in an orderly fashion and in a form readily understood by others: ideas may be more readily extracted, and conclusions more readily drawn from data which have been organized into a coherent pattern.

All observations are affected by variables. For example, the measured rate of a chemical change may be affected by changes in temperature or in the concentrations of the reacting species, or by the presence of light or catalysts, and so on. Before one can proceed very far towards the understanding of an experimental system, recognition of the important variables, at least, is necessary. Science demands the ability to reproduce the results of an experiment, and attempts to do so will only be successful in so far as the important variables are maintained in a similar condition of balance when the experiment is 'repeated'. Confidence in the scientific accuracy of observations comes from this ability to repeat an experiment successfully, and much quantitative data of a repetitive nature originate in this way. These are classified as *univariate* data.

Mere recognition of variables, however, is of limited value, and increased understanding of an experimental system often comes from observing changes in the behavioural pattern of the system when one or more of the variables is altered in a deliberate manner. Thus, for a particular chemical system, it may be noticed that the measured rate of chemical change is doubled when the concentration of one of the reacting substances is doubled. This suggests that, for this system, the rate divided by the concentration may give a value which remained sensibly constant over a range of concentrations. Ability to reproduce this value for different concentrations adds confidence to the suggestion, and leads to a better appreciation of the behavioural nature of the system. Interpretative procedures of this kind provide a second source of repetitive or univariate data.

1–1 Condensation of data

Large amounts of repetitive data may be awkward and unwieldy to use as they stand, and means are sought to condense them to a reasonable compass. All experimental data show some degree of scatter, and a first convenient step is to arrange them in order of magnitude—a procedure known as *ranking*. Condensation can then be effected by arranging the data in classes, that is by listing the frequency (or number of times) with which recorded data lie between specified values or limits. An example of this procedure is shown in Table 1–1.

Table 1–1   Classifying data
Listing the frequencies with which data have been recorded between specified limiting values.

Specified values (class)	Number of data (frequency)
0·7 to 0·8	2
0·8 to 0·9	8
0·9 to 1·0	23
1·0 to 1·1	42
1·1 to 1·2	38
1·2 to 1·3	33
1·3 to 1·4	24
1·4 to 1·5	15
1·5 to 1·6	5
1·6 to 1·7	1
Total number of recordings =	191

Data classified in this way are often graphed in the form of a histogram, as illustrated in Fig. 1–1. In such a figure, the height of each rectangle represents the frequency of the recording of values within the limits shown at the base of the rectangle. The width of the base of each rectangle is known as the *class width*.

Examination of either Table 1–1 or Fig. 1–1, shows that while recordings have been made most frequently between the values 1·0 and 1·1, the *median* value, that is the middle item in the ranking sequence, lies between 1·1 and 1·2. Table 1–1 shows a total of 191 recordings; therefore the median, in this case, is the 96th largest value.

The histogram acts as a pointer to the character of the recorded data. If the number of recorded data were doubled and the class width reduced from 0·1 to 0·05, a more accurate histogram, containing twice the number of steps, could be constructed. It is clear that with a very large number of data and very small class widths the step-wise nature of the histogram would diminish and, in the limit, its outline would approach that of a continuous curve such as the one shown in dashed outline in Fig. 1–1. Thus curve is called a *frequency distribution curve*. Its peak corresponds to the most commonly recorded value, the *mode*. The relative positions of the median and the arithmetic mean values, for the data displayed in Fig. 1–1, are also shown for comparison on this curve.

A useful variation in the representation of the character of the data shown in Fig. 1–1 is to express the vertical axis as a measure of probability, rather than frequency. For example, the first class in Table 1–1 has a frequency of 2, and, with a total of 191 datum values, this corresponds to a probability of $2/191$. When the vertical axis of Fig. 1–1 is expressed as probability, the sum of the heights of all rectangles in the histogram equals unity, corresponding to unit probability of inclusion of all data.

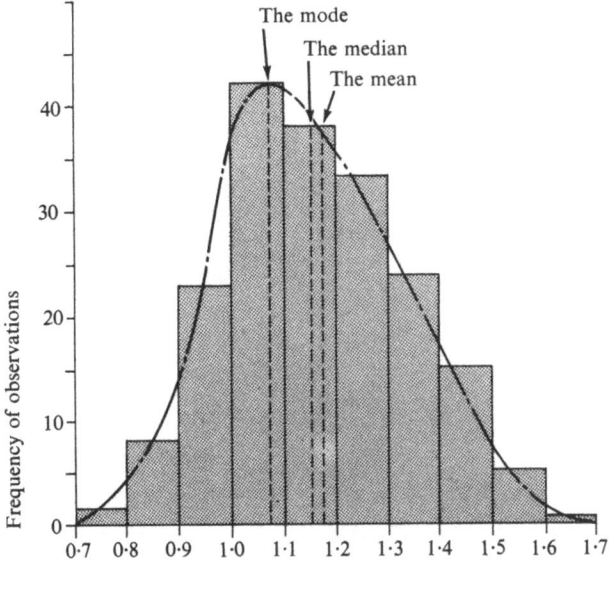

Fig. 1–1 Histogram showing frequency of recorded data from Table 1–1. The frequency distribution curve is shown in dashed (----) outline.

Examination of the shape of the particular frequency distribution curve displayed in Fig. 1–1 shows that it is clearly asymmetric or *skewed*: more than half of the recorded number of data are of magnitude greater than the mode. Other cases of recorded data may arise in which there is no significant skew, or in which the skew is in the opposite sense to that shown in Fig. 1–1 (that is, more than half of the data are of magnitude less than the mode). Very occasionally distribution curves with two maxima (bimodal curves) are encountered. Large numbers of data are required for the accurate identification of the shape of a distribution curve, particularly if it is believed to be of an unusual type.

Two parameters are commonly used to characterize a sample of univariate data. The first of these is a measure of central tendency, corresponding to the most representative value of the sample. The second is a measure of dispersion describing the spread of the data about the central position. These two parameters suffice for most practical purposes, but a measure of skewness is sometimes included in cases of significantly asymmetric distributions; this measure will not be discussed here.

Although the median and the mode also represent measures of central tendency, the arithmetic mean, $\bar{x}$, is usually taken as the most representative value of a sample of data. For n_T values with magnitudes $x_1, x_2, \ldots$, the mean is defined by

$$\bar{x} = \frac{x_1 + x_2 + \cdots}{n_T} = \sum \frac{x_n}{n_T} \tag{1-1}$$

The mean corresponds to the point on the x scale for which the sum of all deviations, $\sum(\bar{x} - x_n)$ equals zero, and for which the sum of the squares of the deviations, $\sum(\bar{x} - x_n)^2$, is a minimum.

The simplest measure of spread is the *range*, which is the difference between the smallest and highest recorded datum values; it is of practical use only for small samples. The *standard deviation* (SD) represents the most useful measure of dispersion and is defined in terms of its square as follows:

$$SD^2 = \frac{\sum(x_n - \bar{x})^2}{n_T - 1} \tag{1-2}$$

The denominator, $n_T - 1$, corresponds to the number of degrees of freedom on which the estimate of the standard deviation is based. It should be understood that, although there are n_T individual values in the sample having a share in the determination of $\bar{x}$, the number of independent differences between these values is $n_T - 1$. Thus, at least two datum values are required before the first estimate of spread can be made.

With large samples, calculations are simplified by the adoption of an assumed mean, x_a, a rounded value of x chosen near the centre of the sample. Its relationship to $\bar{x}$ is given by

$$\bar{x} = x_a + \frac{\sum(x_n - x_a)}{n_T} \tag{1-3}$$

Similarly, the square of the standard deviation is also related to the assumed mean through the following equation:

$$SD^2 = \frac{\sum(x_n - x_a)^2 - n_T(x_a - \bar{x})^2}{n_T - 1} \tag{1-4}$$

In Eqns (1-1) to (1-4), equal weighting is given to each datum value. When $x_1, x_2, \ldots$, have differing reliabilities they may be given different weightings $f_1, f_2, \ldots$, and the weighted mean, $\bar{x}_w$, is then given by Eqn (1-5).

$$\bar{x}_w = \frac{f_1 x_1 + f_2 x_2 + \cdots}{f_1 + f_2 + \cdots} \tag{1-5}$$

Similarly if the sum, $f_1 + f_2 + \cdots$, equals the total number of data, the square of the standard deviation of the weighted data is given by the following equation:

$$SD_w^2 = \frac{\sum f_n (x_n - \bar{x}_w)^2}{(\sum f_n) - 1} \tag{1-6}$$

Alternatively these two formulae may be used for large numbers of data, in which case $f_1, f_2, \ldots$, represent the frequencies in the various classes for which $x_1, x_2, \ldots$, correspondingly represent the class means or median values.

1–2 Reference distribution curves

It is convenient to relate the distribution of data in a sample to some reference or standard distribution curve. This procedure can lead to a better understanding of the manner in which the data is spread around a central value; it can also assist in the making of predictions and in the comparing of different samples of related data.

Of basic importance are the curves generated by the relationship,

$$y = a \exp\left[-b^2 (x - m)^2\right] \tag{1-7}$$

These bell-shaped curves illustrated in Fig. 1–2 are symmetrically disposed with respect to the line $x = m$, at which y attains its maximum value of a. As x departs in value from m, y falls towards zero value, the rate at which it approaches this value being determined by the magnitude of b. It is apparent that these curves could be used to describe the distribution of data about a central value m, the spread of the data being related in an inverse sense to b.

Equation (1–7) may be normalized so that the area under the curve is unity, corresponding to unit probability (of inclusion of all data). In this form, given by Eqn (1–8), the curve is known as the *normal* or *gaussian* (*probability*) *distribution curve*.

$$y = \frac{1}{\sigma\sqrt{(2\pi)}} \exp\left[\frac{-(x - m)^2}{2\sigma^2}\right] \tag{1-8}$$

In this curve, shown in Fig. 1–3, values of y correspond to probability rather than frequency.

The spread of the curve about $x = m$ is determined by the magnitude of σ, large values of σ giving large spreads. Inflexional points on the curve occur at $x = m \pm \sigma$, and the area under that part of the curve limited by these two points equals 0·6827. The corresponding area limited by

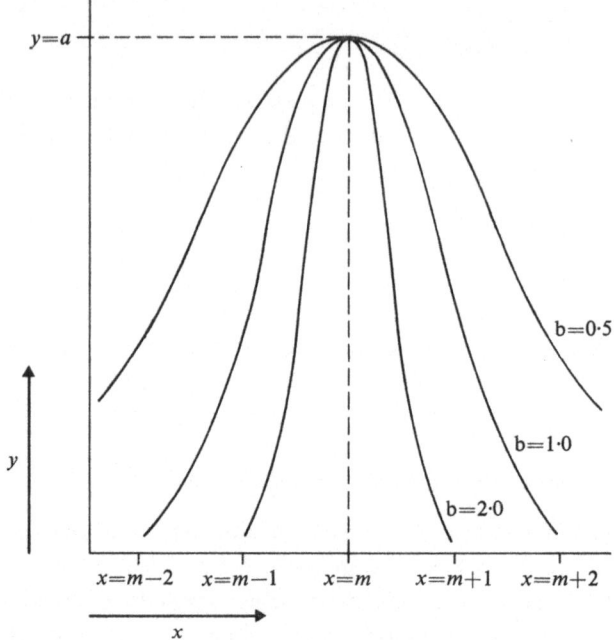

Fig. 1–2   Graphs of $y = a\exp[-b^2(x-m)^2]$ for different values of b.

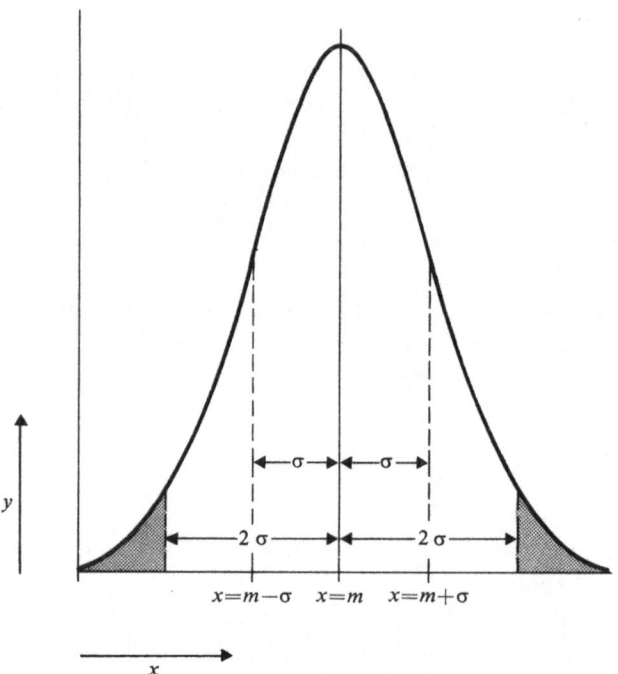

Fig. 1–3   The gaussian distribution curve, $y = [1/\sigma\sqrt{(2\pi)}]\exp[-(x-m)^2/2\sigma^2]$. The shaded area outside the 2σ limits is $4\cdot57\%$ of the total area under the curve.

$x = m \pm 2\sigma$ is 0·9543, and by $x = m \pm 3\sigma$ it is 0·9973. Further details of these areas are given in Table 1–2.

Table 1–2   Areas under the gaussian curve
The area under the gaussian probability curve between the limits $x = m + u\sigma$ and $x = m - u\sigma$.

Area	u	Area	u
0·50	0·67	0·975	2·24
0·60	0·84	0·980	2·33
0·70	1·04	0·985	2·43
0·80	1·28	0·990	2·58
0·85	1·44	0·991	2·61
0·90	1·64	0·992	2·65
0·91	1·69	0·993	2·70
0·92	1·75	0·994	2·75
0·93	1·81	0·995	2·81
0·94	1·87	0·996	2·88
0·95	1·96	0·997	2·97
0·96	2·05	0·998	3·09
0·97	2·17	0·999	3·29

For data conforming precisely to the distribution curve given by Eqn (1–8), the mean ($\bar{x}$) corresponds to m, and the standard deviation corresponds to σ. An amount 68·27% of the data lies between $m + \sigma$ and $m - \sigma$, 95·43% lies between $m + 2\sigma$ and $m - 2\sigma$, and so on.

To check the conformity of data to a gaussian distribution, one would require a very large sample. (To be quite sure, the sample would need to be of infinite size!) The limitations imposed by the experimental approach often confine one in practice to quite small samples. Even if these samples were drawn randomly from an infinite 'population' of truly gaussian character, the sample characteristics would not be precisely the same as those of the infinite population, and would vary from sample to sample. Thus, the mean and the standard deviation of one sample would be slightly different from those of another sample. For an infinite number of such samples of size n_T drawn randomly from an infinite population of this type with standard deviation σ, the means of the samples are also distributed in a gaussian pattern (that is, they are normally distributed) with a standard deviation of $\sigma/\sqrt{n_T}$. This value is called the *standard error of the mean* (SEM).

$$\text{SEM} = \frac{\sigma}{\sqrt{n_T}} \tag{1–9}$$

For a sample of finite size n_T, which is assumed to be drawn from a gaussian population, the standard deviation provides an estimate of σ, and $\text{SD}/\sqrt{n_T}$ provides an estimate of the standard error of the mean. The larger the sample, the more certain are these estimates. Of the area under a gaussian curve, 95·43% lies between $m+2\sigma$ and $m-2\sigma$; judging, therefore, from a sample of 100 or more data, it could be said with reasonable certainty that 95% of the data in the whole (hypothetically infinite) population would lie between $\bar{x}+2SD$ and $\bar{x}-2SD$. Also, one could reasonably predict that the means of other samples of similar size would have 95% probability of lying between $\bar{x}+2SD/\sqrt{n_T}$ and $\bar{x}-2SD/\sqrt{n_T}$, these values having been estimated from the original sample. Such predictions would be made with considerably less certainty from a sample of ten data. The usefulness of the standard deviation as a predictor decreases considerably as the sample size decreases, and one must be much more cautious when making predictions from small samples.

Some compensation for errors involved in predicting from a small sample is provided by the use of so-called t values, listed in Table 1–3. From a sample of n_T data it may be predicted with some confidence that

Table 1–3 Table of t values
The t values are listed for different degrees of freedom (ϕ) and limits of inclusion $(1-P)$.

		t values	
	for $P = 0.1$	for $P = 0.05$	for $P = 0.01$
$\phi = $ 1	6·31	12·71	63·66
2	2·92	4·30	9·93
3	2·35	3·18	5·84
4	2·13	2·78	4·60
5	2·02	2·57	4·03
6	1·94	2·45	3·71
7	1·90	2·37	3·50
8	1·86	2·31	3·36
9	1·83	2·26	3·25
10	1·81	2·23	3·17
11	1·80	2·20	3·11
12	1·78	2·18	3·06
13	1·77	2·16	3·01
14	1·76	2·15	2·98
15	1·75	2·13	2·95
16	1·75	2·12	2·92
17	1·74	2·11	2·90
18	1·73	2·10	2·88
19	1·73	2·09	2·86
20	1·73	2·09	2·85
25	1·71	2·06	2·79
30	1·70	2·04	2·75
40	1·68	2·02	2·70
50	1·68	2·01	2·68
60	1·67	2·00	2·66
80	1·66	1·99	2·64
100	1·66	1·98	2·63

95% of the data in the whole population (assumed gaussian) will be between the limits, $\bar{x} \pm t \times SD$, t being taken for 95% limits of inclusion ($P = 0.05$) and $n_T - 1$ degrees of freedom (ϕ). These limits, determined with the aid of t values are called *confidence limits*. Similarly,

$$\text{confidence limits of the mean} = \pm t \times \frac{SD}{\sqrt{n_T}} \qquad (1\text{–}10)$$

Reference to the table shows that the factor, t, for 95% limits of inclusion is 2.26 for $n_T = 10$ ($\phi = 9$), and 1.98 (close to 2) for $n_T = 100$ ($\phi = 99$). This table also lists corresponding t values for 90% limits ($P = 0.1$) and 99% limits ($P = 0.01$).

The association of particular confidence limits with an estimated mean value implies a prediction about the reliability of this estimate of the mean, as judged from both the spread of the data and the number of data in the sample from which the mean has been estimated. For example, when 90% confidence limits of the mean are given for a particular sample of data, it is being predicted that, if repeated attempts were made to reproduce this sample of data from many further experiments, then, on an average of nine out of ten occasions, the mean of each new sample would lie within the 90% confidence limits of the mean of the initial sample.

It is clear that the gaussian distribution curve represents a most useful mathematical model, and, because of this, it is customary to apply it to experimental data, even when it is suspected that the data show some deviation from normal or gaussian behaviour. The mean and its confidence limits provide a convenient pair of parameters for the coding of univariate data in terms of a measure of central tendency and its reliability (90% confidence limits are commonly used). One inconsistency which appears to be commonly tolerated with kinetic data is to assume gaussian behaviour both for recorded values of the rate coefficient, k, at a given temperature, and for values of log k when they are expressed as a function of temperature in terms of the Arrhenius equation, $\log k = \log A - E/2.303RT$. Clearly, k and log k cannot be both normally distributed at the one time. Further discussion of this point is made in Chapter 2.

The conformity of data to gaussian behaviour can be checked qualitatively by the use of probability graph paper or more precisely by means of the 'chi-squared test'. The latter test is not discussed here: interested readers may seek further information about it from the references at the end of this chapter. With probability graph paper, the cumulative probabilities of the occurrence of data up to the limit of chosen values are plotted against these values. A nonlinear scale is used for probabilities, so that gaussian data plot as a straight line. The use of probability graph paper is illustrated in Fig. 1–4 for the data previously listed in Table 1–1 (Table 1–4 relists this data in terms of cumulative probabilities).

Table 1–4   Testing for gaussian behaviour
Cumulative frequencies and probabilities of data from Table 1–1 having been recorded below specified values.

Specified value	Cumulative frequency	Cumulative probability
0·7	0	0·00
0·8	2	0·01
0·9	10	0·05
1·0	33	0·17
1·1	75	0·39
1·2	113	0·59
1·3	146	0·76
1·4	170	0·89
1·5	185	0·97
1·6	190	0·99
1·7	191	1·00

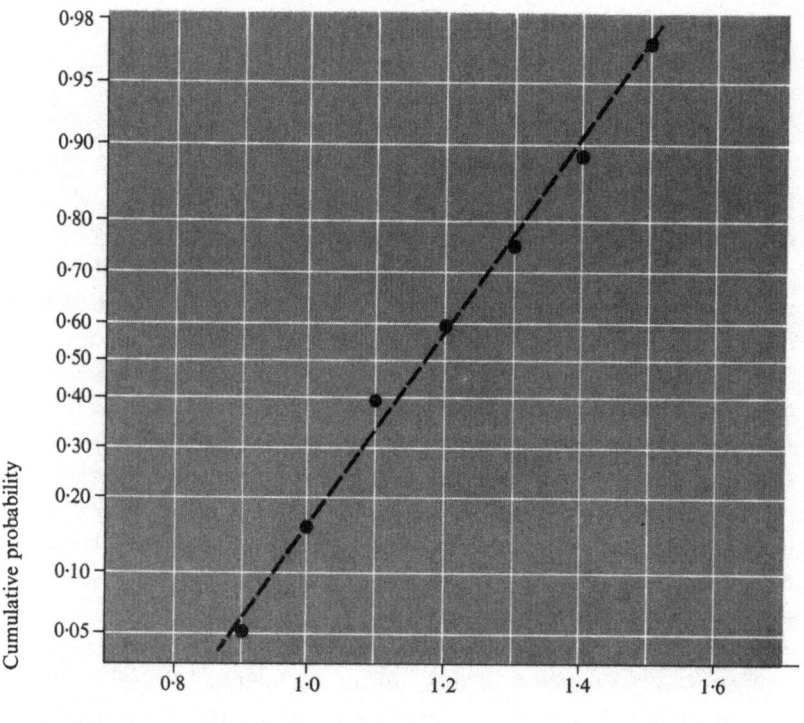

Fig. 1–4   Testing for gaussian behaviour. Data from Table 1–1 plotted on probability graph paper.

Although discussion so far has been restricted to the gaussian distribution because of its almost universal use, there are many others, such as the binomial, the Cauchy, and the Poisson distributions. The last named merits some consideration here because of its application in certain chemical kinetic systems.

The Poisson distribution is particularly appropriate to the analysis of the occurrence of isolated events in a measurement continuum. Examples of this nature are the measurement of dust particles in the atmosphere (isolated particles in a continuum of space), and the emission of subatomic particles from a sample undergoing radiochemical change (isolated events in a continuum of time). It may also apply to induction periods and explosions in chemical kinetic systems.

According to the Poisson model, the probability, P, of occurrence of f events within a specified amount of the continuum is given by

$$P = m^f [\exp(-m)]/f! \tag{1–11}$$

when m is the average number of occurrences within this amount of continuum. The sum of all probabilities, P, must be unity; therefore, summing for $f = 0, 1, 2, …, \infty$,

$$\sum P = 1 = [1 + m + m^2/2! + \cdots] \exp(-m) \tag{1–12}$$

In Eqn (1–12), the summation inside the squared brackets equals $\exp(+m)$, for values of f to infinity, and this leads to the required value of unity when it is multiplied by the term $\exp(-m)$, outside the squared brackets.

Consider the application of this probability distribution to the emission of α-particles from a radioactive source: if the average number of particles emitted per second is 4·03, the Poisson model predicts that the probability of only two particles being emitted in any one-second interval is,

$$P_2 = 4·03^2 [\exp(-4·03)]/2!$$

$$= 0·14$$

Therefore, this is predicted as occurring 14 times in every 100 s.

1–3   Tests of significance and rejection of data

Circumstances often arise in which judgements need to be made about different samples of related data. As an example, one may wish to compare measurements on the rate of decomposition of a pure substance with those corresponding to the substance decomposing in the presence of a suspected contaminant. Typical questions which arise are: (a) 'Is there any significant difference between the two sets of measurements?', or (b) 'What is the likelihood of the apparent difference between the two sets occurring purely by chance?' These questions are also pertinent to the case of a single result being judged against a set of results.

The use of a standard model, such as the gaussian distribution, places these questions and their answers within a more quantitative framework, but the final criteria upon which judgements are made must still be determined by the observer. The arbitrary assumption is commonly made that results differ significantly when the probability of the difference arising by chance is less than one in twenty. More critical observers, however, may wish to set the significance level to one in a hundred, and less critical observers to one in ten.

Suppose, for example, that, from ten kinetic runs on the decomposition of a pure substance, the average value of the first-order rate coefficient, k, is found to be 2.21×10^{-5} s^{-1} with a standard deviation of 0.11×10^{-5} s^{-1}. Table 1–3 shows that the t value is 2·26 for nine degrees of freedom and 95% limits ($P = 0.05$); the 95% confidence limits of the data are therefore $\pm 2.26 \times 0.11 \times 10^{-5}$ s^{-1}, that is, $\pm 0.25 \times 10^{-5}$ s^{-1}. According to this estimate, then, 95 out of 100 kinetic runs on the pure substance will have a k value lying in the range $(1.96 - 2.46) \times 10^{-5}$ s^{-1}. A value lying outside this range will have less than a one in twenty chance of belonging to the population from which this sample of ten kinetic runs was drawn, and, adopting the one in twenty criterion, it is judged to differ significantly from this sample. Similar arguments may be made in terms of the one in ten, or one in a hundred criterion.

When comparing two samples of data, arguments are more appropriately based on the confidence limits of the means of the samples. The mean of the sample just referred to has an estimated standard error of 0.034×10^{-5} s^{-1} and 95% confidence limits of $\pm 0.08 \times 10^{-5}$ s^{-1}. Any other sample with similar confidence limits, but having a mean lying outside the range $(2.13 - 2.29) \times 10^{-5}$ s^{-1} may be judged, on the one in twenty criterion, to differ significantly from the first sample. If the second sample of data corresponded to the decomposition of the substance in the presence of a suspected contaminant, the contaminant would be judged to have a significant effect upon the rate of decomposition of the substance. Depending upon one's understanding of the chemical nature of the system, however, the effect, although significant, may or may not be judged as being chemically important.

The test just described is a 'two-sided test', that is, the value being tested is judged against both ends of the distribution curve associated with the first sample (see Fig. 1–3). The two-sided test should be applied when there is no definite reason to account for the difference between the two samples being of a particular sign. If, for example, the possible contaminant were known to be a catalyst, then a one-sided test would be justified, because one would be testing for a significant positive increase in decomposition rate in the presence of the contaminant. The chance of a value lying under only one end of the distribution curve is, of course, one-half of its chance of lying under either end.

It is clear that these tests could also be used for rejecting suspected data from a sample, but the one in twenty criterion is probably too strict for very small samples and not sufficiently strict for large samples. A reasonable compromise is to reject values from a sample of size n_T when they are judged to have a probability of occurrence of less than $1/2n_T$, this fraction being limited to a maximum of $1/10$ (Table 1–2 is useful for tests such as this). It has been argued that the suspected value should be excluded when calculating the sample mean, but included when calculating confidence limits, however this approach is probably too restrictive for small samples.

The rejection of suspected data from a sample is always a risky process, and difficult to carry out without prejudice. If a value is under suspicion because of some doubt about the experiment from which it was obtained, then it should be rejected whether or not it 'looks right'. Decisions of this type should not be unduly coloured by an unnatural, but understandable, desire for the tidiness of data.

Examples and problems

Example 1–1

Calculation of standard deviation and 90% confidence limits of the mean from the following experimental results for the measured rate coefficient of a chemical reaction:

$$10^5 k \, (s^{-1}) = 16{\cdot}7; \; 17{\cdot}0; \; 17{\cdot}1; \; 17{\cdot}2; \; 17{\cdot}2; \; 17{\cdot}4; \; 17{\cdot}6; \; 18{\cdot}0 \quad (x_n \text{ values})$$

$$n_T = 8; \qquad \phi = n_T - 1 = 7$$

$$\bar{x} = \sum x_n/n_T \quad \text{Eqn (1–1)}$$

$$= 17{\cdot}28$$

$$SD^2 = [\sum (x_n - \bar{x})^2]/(n_T - 1) \quad \text{Eqn (1–2)}$$

$$= [0{\cdot}58^2 + 0{\cdot}28^2 + \cdots]/7$$

$$= 0{\cdot}156$$

Alternative method using an assumed mean, $x_a = 17{\cdot}2$:

$$\bar{x} = x_a + \sum (x_n - x_a)/n_T \quad \text{Eqn (1–3)}$$

$$= 17{\cdot}2 + (-0{\cdot}5 - 0{\cdot}2 - 0{\cdot}1 + 0 + 0 + 0{\cdot}2 + 0{\cdot}4 + 0{\cdot}8)/8$$

$$= 17{\cdot}2 + 0{\cdot}08 = 17{\cdot}28$$

$$SD^2 = [\sum (x_n - x_a)^2 - n_T(x_a - \bar{x})^2]/(n_T - 1) \quad \text{Eqn (1–4)}$$

$$= [(0{\cdot}25 + 0{\cdot}04 + 0{\cdot}01 + \cdots) - 8(0{\cdot}08)^2]/7$$

$$= [1{\cdot}14 - 0{\cdot}05]/7$$

$$= 0.156$$

$$SD = 0.40$$

Confidence limits of the mean $= \pm t \times SD/\sqrt{n_T}$ Eqn (1–10)

t for 90% limits and $\phi = 7$ is 1·90   (Table 1–3)

90% confidence limits of the mean $= \pm 1.90 \times 0.40/\sqrt{8}$

$$= \pm 0.27$$

Result: $10^5 \bar{k}\ (s^{-1}) = 17.28 \pm 0.27$ (90% confidence limits of the mean).

Problems

1–1   The following results were reported for the measured rate coefficient of a chemical reaction:

$10^5 k\ (s^{-1}) = 89.5;\ 90.6;\ 91.3;\ 91.6;\ 91.9;\ 92.0;\ 92.2;\ 92.5;\ 92.7;\ 93.0;\ 93.6;$
$93.9;\ 94.1;\ 94.8;\ 95.0$

(a) Estimate the mean, the standard deviation, and the 90% confidence limits of the mean of these results.

(b) A single further result, $10^5 k\ (s^{-1}) = 86.0$, was reported. Use a statistical test to decide whether or not this result should be rejected.

1–2   The table given below lists counts of α-particles from a radioactive source as measured over a succession of 10 s periods. Estimate the mean number of counts per period, and check the distribution of counts as predicted by the Poisson model against those actually observed.

Counts per 10 s period	Number of periods
0	60
1	201
2	385
3	521
4	535
5	409
6	269
7	142
8	46
9	25
10	11

1–3   (a) Construct histograms based on the Poisson distributions for,

(i) $m = 1; f = 0, 1, 2, 3, 4$
(ii) $m = 2; f = 0, 1, 2, 3, 4, 5$
(iii) $m = 5; f = 0, 1, 2, 3, 4, 5, 6, 7, 8, 9, 10$

(b) Plot the data for (a)(iii) on probability paper designed to test for gaussian behaviour (see Fig. 1–4). Do you conclude that the gaussian model may be reasonably used in place of the Poisson model in this instance?

1–4   The rate of a gas-phase reaction was followed in a static system by noting the increase in pressure with time. The first-order rate coefficient for the reaction was estimated by use of the integrated form of the rate equation and by Guggenheim's method (see pp. 72 and 79), the two sets of values being listed in Table 1–5, as k_i and k_g respectively. It is suspected that, because the system has an appreciable dead-space (see p. 92), the values for k_g are significantly greater than for k_i. Test this by determining whether the ratio k_g/k_i is significantly greater than unity.

1–5   Table 1–6 lists Arrhenius parameters obtained from a number of independent studies of the first-order pyrolyses of monochloroalkanes in the gaseous phase (A. Maccoll, Chem. Rev., 1969, **69**, 40).

(a) Estimate the mean, and the 95% confidence limits of the mean for
  (i) the values of log A for the primary compounds;
  (ii) the values of log A for the secondary compounds;
  (iii) the values of E for the primary compounds;
  (iv) the values of E for the secondary compounds.
(b) Do the values of log A obtained for the primary compounds differ significantly from those obtained for the secondary compounds?
(c) Do the values of E obtained for the primary compounds differ significantly from those obtained for the secondary compounds?

Table 1–5 Rate coefficient for a gas-phase reaction

(See Problem 1–4)
t = temperature °C; k_i = rate coefficient (s^{-1}) from the integrated form of the rate equation; k_g = rate coefficient by Guggenheim method.

t	$10^5 \times k_i$	$10^5 \times k_g$
334	5·94	5·95
334	5·82	6·62
334	6·19	7·34
342	10·24	10·81
342	10·18	9·63
343	10·09	11·78
342	9·27	11·09
350	17·2	17·3
351	18·0	17·1
350	15·4	16·0
368	51·0	54·1
368	51·3	55·3
368	51·4	56·6
368	52·1	52·9
368	51·6	52·2
377	88·9	95·4
377	90·2	93·7
377	88·6	90·5
377	91·8	93·9
377	88·9	92·2
377	89·8	93·5

Table 1–6 Arrhenius parameters for mono-chloroalkane pyrolyses

(See Problem 1–5)

A in s^{-1}; E in kcal mole^{-1}.

	$\log A$	E
Primary		
C_2H_5Cl	13·16	56·4
	13·46	56·6
	14·03	58·4
	13·51	56·6
$n\text{-}C_3H_7Cl$	13·45	55·0
	13·50	55·1
$n\text{-}C_4H_9Cl$	14·50	57·9
	14·00	57·0
	13·63	55·2
$n\text{-}C_5H_{11}Cl$	14·61	58·3
$i\text{-}C_4H_9Cl$	14·02	56·9
	13·81	55·3
$n\text{-}C_{10}H_{21}Cl$	14·1	55·7
Secondary		
$i\text{-}C_3H_7Cl$	13·40	50·5
	13·64	51·1
$s\text{-}C_4H_9Cl$	13·62	49·6
	14·00	50·6
	14·07	50·8
$c\text{-}C_5H_9Cl$	13·47	48·3
$c\text{-}C_6H_{11}Cl$	13·77	50·0
	13·88	50·2
$s\text{-}C_8H_{17}Cl$	13·53	48·7

(Table abstracted from A. Maccoll, *Chem. Rev.*, 1969, **69**, 40.)

References

1. Moroney, M. J. *Facts from Figures.* Penguin, London, 1953.
2. Topping, J. *Errors of Observation and Their Treatment.* Institute of Physics, London, 1955.
3. Lark, P. D., B. R. Craven, and R. C. L. Bosworth. *The Handling of Chemical Data.* Pergamon Press, Oxford, 1968.
4. Youden, W. J. *Statistical Methods for Chemists.* John Wiley, New York, 1951.
5. Mandel, J. *The Statistical Analysis of Experimental Data.* John Wiley (Interscience), New York, 1964.
6. Parratt, L. G. *Probability and Experimental Errors in Science.* John Wiley, New York, 1961.

2 Observing change

When observations are made upon a system undergoing change, the observer is commonly led to suspect that there are direct associations between certain of the variables in the system. Attempts to represent these associations quantitatively assist in the classification of the behavioural pattern of the system, and may lead to an interpretation of this behaviour in terms of a mathematical or physical model. In the case of a chemical kinetic system this model could correspond to a mechanism for the chemical change.

When the association exists among a number of variables, the nature of the interdependence between a pair of these may be more difficult to evaluate, particularly if control of variables other than these two is hard to accomplish. In many cases, however, it is possible to maintain the other variables in a reasonably constant or well-controlled condition, while measurements are made upon the pair of interest to the observer.

Data obtained from this type of activity are usually represented in the form of tables, graphs, or mathematical equations. Equations represent the ultimate in condensation of data, but they are, of course, highly interpretative; graphs are convenient as regards ease of reference, and are useful for displaying trends, but they are of limited value for quantitative work. Since original data are usually in tabular form, it is important to have an understanding of the range of general methods available for extracting information directly from tables. It is the author's experience that the value and the range of these methods are often not sufficiently well appreciated, and they therefore merit some discussion at this point. Special methods, more particularly related to chemical kinetic systems, are discussed in the later chapters.

2–1 Nature of tabulated data. Item differences

When associated observations are made on a pair of variables, such as *time* and *concentration*, it is usually convenient to assume one of these to be independent (for example *time*), and the other to be dependent. Measured values of the independent variable are taken as being precise, and scatter of the data is then tied to measurements of the dependent variable. Although the choice of dependent and independent variable is often made on reasonably realistic grounds, in many instances it is assumed only as a convenient mathematical fiction. Occasionally, analysis is made on the assumption that both variables are subject to error.

Tables of associated data generally list values of the independent variable in order of increasing or decreasing magnitude, the difference between successive values being called *item differences*, or *item intervals*. Particularly useful tables are those which list the independent variable in rounded values and which have the item intervals constant. Raw experiment data are rarely of this form, but it is often worth while designing experiments in such a way that the results may be cast into this form by minor use of interpolation methods. Data so arranged may be organized into the form of an item-difference table of the type shown in Table 2–1.

Table 2–1   Item difference table, with constant x intervals

x	y	Δy	$\Delta^2 y$	$\Delta^3 y$
x_0	y_0			
		Δy_0		
$x_1 = x_0 + \Delta x$	y_1		$\Delta^2 y_0$	
		Δy_1		$\Delta^3 y_0$
$x_2 = x_0 + 2\Delta x$	y_2		$\Delta^2 y_1$	
		Δy_2		$\Delta^3 y_1$
$x_3 = x_0 + 3\Delta x$	y_3		$\Delta^2 y_2$	
...	...	...	...	...
$x_n = x_0 + n \Delta x$	y_n	...	...	...

In Table 2–1, x represents the independent variable and y the dependent variable. Values of Δy, $\Delta^2 y$, and $\Delta^3 y$ are called respectively the first-order, second-order, and third-order differences in y. They are related to the y values by equations such as the following:

$$\Delta y_0 = y_1 - y_0 \tag{2–1}$$

$$\Delta y_1 = y_2 - y_1 \tag{2–2}$$

$$\Delta^2 y_0 = \Delta y_1 - \Delta y_0 = y_2 - 2y_1 + y_0 \tag{2–3}$$

$$\Delta^3 y_0 = \Delta^2 y_1 - \Delta^2 y_0 = y_3 - 3y_2 + 3y_1 - y_0 \tag{2–4}$$

The differences, Δy, $\Delta^2 y$, $\Delta^3 y$ are related in turn to the first-order, second-order, and third-order differential coefficients, dy/dx, $d^2 y/dx^2$, $d^3 y/dx^3$. For this reason, the item-difference table can provide an indication of the limiting degree of the polynomial equation of the general form,

$$y = a + bx + cx^2 + \cdots \tag{2–5}$$

which may be appropriate for the data. When a first-degree equation is appropriate, values of Δx in the table are approximately constant; with a second-degree equation, values of $\Delta^2 x$ are approximately constant, and so on. In these cases, the higher order differences show no significant net drift in magnitude, and fluctuate around zero. Application of this *tabular suitability test* to the data listed in Table 2–2, demonstrates that, in this example, a quadratic equation is appropriate. Assuming that $\Delta^2 y/(\Delta x)^2 \approx d^2 y/dx^2$, the coefficient of the squared term in this quadratic equation is given approximately by $c \approx \Delta^2 y/2(\Delta x)^2 \approx -2$.

Table 2–2   Illustrating tabular differences for recorded data

x	y	Δy	$\Delta^2 y$	$\Delta^3 y$	$\Delta^4 y$
1·0	3·01				
		0·43			
1·1	3·44		−0·03		
		0·40		−0·02	
1·2	3·84		−0·05		+0·05
		0·35		+0·03	
1·3	4·19		−0·02		−0·06
		0·33		−0·03	
1·4	4·52		−0·05		+0·05
		0·28		+0·02	
1·5	4·80		−0·03		−0·05
		0·25		−0·03	
1·6	5·05		−0·06		+0·06
		0·19		+0·03	
1·7	5·24		−0·03		
		0·16			
1·8	5·40				

Item-difference tables can also give a valuable guide to the smoothness of data, for lack of regularity in the change of y values down the table will be reflected by considerably greater disturbances in the succeeding values of Δy, $\Delta^2 y$, and $\Delta^3 y$. As a demonstration of this, Table 2–3 lists, for various times, the concentration of a substance which is decomposing according to a strict second-order kinetic law, $-dC/dt = kC^2$; it may be observed that the differences ΔC, $\Delta^2 C$, etc. change in a smooth, regular manner down the table. In Table 2–4, the value for the concentration at time, $t = 50$ s has been deliberately changed so that it is in error; the corresponding disturbances in the succeeding first-, second-, and third-order item differences in C are quite apparent, and it is also clear that these disturbances are at a maximum along the row of difference values in line with the incorrect C value. Tables such as this therefore provide useful indicators for possible errors in recorded data.

Where data show overall scatter, smoothing may be effected by the use of modified differences obtained from 'curves of best fit' of raw Δx or $\Delta^2 x$ values plotted against y. An illustration of this procedure is shown at the end of the chapter (Example 2–1). The excessive use of smoothing methods is always open to criticism, but smoothing is justified

Table 2–3 Decomposition by a second-order law
t = time from start of chemical reaction (seconds); C = concentration of decomposing substance (millimoles per litre).

t	C	ΔC	$\Delta^2 C$	$\Delta^3 C$	$\Delta^4 C$
0	51·0				
		−4·6			
10	46·4		+0·7		
		−3·9		−0·1	
20	42·5		+0·6		0·0
		−3·3		−0·1	
30	39·2		+0·5		0·0
		−2·8		−0·1	
40	36·4		+0·4		0·0
		−2·4		−0·1	
50	34·0		+0·3		0·0
		−2·1		−0·1	
60	31·9		+0·2		+0·1
		−1·9		0·0	
70	30·0		+0·2		0·0
		−1·7		0·0	
80	28·3		+0·2		0·0
		−1·5		0·0	
90	26·8		+0·2		
		−1·3			
100	25·5				

Table 2–4 Detecting an error with a difference table
Concentration at time, $t = 50$, is in error (data based on Table 2–3).

t	C	ΔC	$\Delta^2 C$	$\Delta^3 C$	$\Delta^4 C$
0	51·0				
		−4·6			
10	46·4		+0·7		
		−3·9		−0·1	
20	42·5		+0·6		0·0
		−3·3		−0·1	
30	39·2		+0·5		+0·3
		−2·8		+0·2	
40	36·4		+0·7		−0·3
		−2·1		−1·0	
50	34·3		−0·3		+1·8
		−2·4		+0·8	
60	31·9		+0·5		−1·1
		−1·9		−0·3	
70	30·0		+0·2		+0·3
		−1·7		0·0	
80	28·3		+0·2		0·0
		−1·5		0·0	
90	26·8		+0·2		
		−1·3			
100	25·5				

where its immediate intent is to set data into a form suitable for subsequent mathematical manipulation: many numerical methods will not operate sensibly on unsmoothed data. Belief in the value of smoothed data is based on the philosophy that all observations are subject to error, and on the experimenter's conviction that the association between variables in the system is a continuous one and not subject to wild fluctuations.

With tables such as 2–3 it is relatively easy to extrapolate to unlisted t values just outside the range of recorded data. An estimate of C for $t = 110$ in this table, for example, may be made by assuming the following subsequent values at the bottom of the table: $\Delta^3 C = 0\cdot0$, $\Delta^2 C = 0\cdot2$, $\Delta C = -1\cdot1$, and therefore $C_{(t=110)} = 24\cdot4$. This technique of extrapolation is extremely useful for kinetic data when it is necessary to estimate the concentration at zero time from subsequent concentration measurements.

It is not always convenient or appropriate to arrange data into a form with constant item differences in x. In these cases the following system of *divided differences* (Δ_d) may be defined:

(a) First divided differences:

$$\Delta_d y_0 = (y_1 - y_0)/(x_1 - x_0) \tag{2–6}$$

$$\Delta_d y_1 = (y_2 - y_1)/(x_2 - x_1) \tag{2–7}$$

$$\Delta_d y_2 = (y_3 - y_2)/(x_3 - x_2) \tag{2–8}$$

and so on. (The first divided differences correspond approximately to the average value of the derivative, dy/dx, over the range of the x interval.)

(b) Second divided differences:

$$\Delta_d^2 y_0 = (\Delta_d y_1 - \Delta_d y_0)/(x_2 - x_0) \tag{2–9}$$

$$\Delta_d^2 y_1 = (\Delta_d y_2 - \Delta_d y_1)/(x_3 - x_1) \tag{2–10}$$

and so on.

(c) Third divided differences:

$$\Delta_d^3 y_0 = (\Delta_d^2 y_1 - \Delta_d^2 y_0)/(x_3 - x_0) \tag{2–11}$$

and so on.

2–2   Interpolation and extrapolation

The estimation of a value y_n corresponding to an unlisted value x_n is called *interpolation* when the unknown value lies within the recorded range of data, and *extrapolation* when it lies outside this range. The methods of interpolation and extrapolation should always be used with caution, particularly extrapolation which assumes a predictability of behaviour beyond the limit of observation. Most of the methods are based on the assumption that the data may be represented in terms of a convergent polynomial relation of the form given in Eqn (2–5). This assumption serves for most kinetic data. However, when the character of data is known to undergo rapid change, the assumption of convergent polynomial representability may not be valid, and in these cases special methods are required: information about these methods should be sought from the references at the end of this chapter.

The simple *linear method* assumes that y varies linearly with x. Thus, the estimate of y_n is related to known values x_1, y_1, and x_2, y_2 by the following relationship:

$$y_n = y_1 + (y_2 - y_1)\frac{x_n - x_1}{x_2 - x_1} \qquad (2\text{–}12)$$

This method is reasonable when the value being sought is close to a recorded value. Some allowance for curvature may be made by averaging a value obtained from linear interpolation with one obtained from linear extrapolation, as shown in Fig. 2–1, but for greater accuracy the *Gregory–Newton* or *Lagrange method* is preferred.

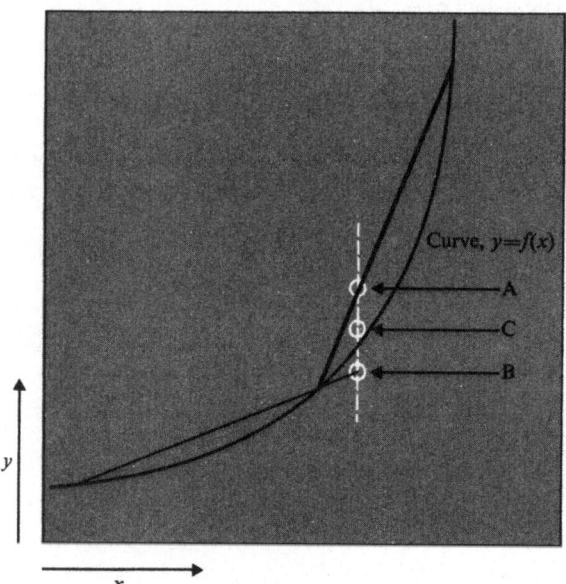

Fig. 2–1 Linear interpolation and extrapolation. (A) Value obtained by linear interpolation; (B) Value obtained by linear extrapolation; (C) Mean of (A) and (B) values.

The Gregory–Newton method is particularly appropriate for smooth data arranged in the form of an item-difference table as illustrated by Table 2–1. With the x values evenly spaced, y_n is related to x_n by the following equation:

$$y_n = y_0 + n\,\Delta y_0 + n(n-1)\frac{\Delta^2 y_0}{2!} + n(n-1)(n-2)\frac{\Delta^3 y_0}{3!} + \cdots \tag{2-13}$$

In this equation,

$$n = (x_n - x_0)/\Delta x$$

The method makes use of tabular data continually advancing from the reference point x_0, y_0 which should be chosen close to the unknown value. It is convenient, therefore, to rewrite a table in the reverse sense when estimating values near the bottom of the table. In any case, when interpolating from the body of a table, it is advisable to check by the use of data advancing in both senses—in this way, maximum use is made of data which straddle the points near which interpolation is required, and slight scatter of data is allowed for.

The Gregory–Newton formula is also expressible in the divided difference form given by Eqn (2–14), this form of the equation being appropriate for cases in which the x values are not evenly spaced.

$$y_n = y_0 + (x_n - x_0)\,\Delta_d y_0 + (x_n - x_0)(x_n - x_1)\,\Delta_d^2 y_0$$
$$+ (x_n - x_0)(x_n - x_1)(x_n - x_2)\,\Delta_d^3 y_0 + \cdots \tag{2-14}$$

As an alternative to the Gregory–Newton method, the Lagrange method is particularly useful for data with unequal x intervals. Equation (2–15) gives the Lagrange formula for r data points:

$$y_n = y_1 \left(\frac{x_n - x_2}{x_1 - x_2} \times \frac{x_n - x_3}{x_1 - x_3} \times \frac{x_n - x_4}{x_1 - x_4} \times \cdots \right)$$
$$+ y_2 \left(\frac{x_n - x_1}{x_2 - x_1} \times \frac{x_n - x_3}{x_2 - x_3} \times \frac{x_n - x_4}{x_2 - x_4} \times \cdots \right)$$
$$+ \cdots \text{ to a total of } r \text{ terms} \tag{2-15}$$

Application of the formula shows that the data near y_n, x_n make the most important contributions to the estimate. Examples illustrating the application of these interpolation and extrapolation methods are given at the end of this chapter (Examples 2–2 to 2–5 inclusive).

2–3 Differentiation and integration

The Gregory–Newton expression (Eqn (2–13)) may be differentiated to yield the following relationship for the derivative, dy/dx, at the point x_n, y_n:

$$\frac{dy}{dx} = \frac{1}{\Delta x} \left[\Delta y_0 + (2n-1)\frac{\Delta^2 y_0}{2} + (3n^2 - 6n + 2)\frac{\Delta^3 y_0}{6} \right.$$
$$\left. + (4n^3 - 18n^2 + 22n - 6)\frac{\Delta^4 y_0}{24} + \cdots \right] \tag{2-16}$$

As defined before, $n = (x_n - x_0)/\Delta x$, and the equation is applicable to tabulated data in which the values of x are equally spaced. When these values are unequally spaced, the divided difference form of the equation may be used:

$$\frac{dy}{dx} = \Delta_d y_0 + [(x_n - x_0) + (x_n - x_1)] \Delta_d^2 y_0$$
$$+ [(x_n - x_0)(x_n - x_1) + (x_n - x_0)(x_n - x_2)$$
$$+ (x_n - x_1)(x_n - x_2)] \Delta_d^3 y_0 + \cdots \qquad (2\text{--}17)$$

Both of these differentiated Gregory–Newton expressions are only applicable to smooth data. Tabular points should be spaced sufficiently far apart to provide reasonable magnitude in the values of Δy.

One special case of Eqn (2–16) is of particular interest. If $n = 1/2$, the term involving $\Delta^2 y_0$ vanishes and

$$\frac{dy}{dx} = \frac{1}{\Delta x} \left[\Delta y_0 - \frac{\Delta^3 y_0}{24} + \cdots \right] \qquad (2\text{--}18)$$

For much experimental data, which necessarily involve some scatter, $\Delta^3 y_0 / 24$ and terms involving differences of higher order are of minor significance. Equation (2–18) therefore demonstrates that the frequently used approximation, $\Delta y / \Delta x \approx dy/dx$, is, in many cases, an accurate approximation at the mid-point of the interval Δx.

For cases in which the basic data are not sufficiently smooth for direct application of the Gregory–Newton formulae, it is usually convenient to graph values of Δy (or $\Delta_d y$) against x and draw a smooth curve through the points so plotted. Smoothed values for the higher order differences may also be derived from this graph, thus reducing annoying fluctuations in these differences when applying Eqns (2–16) and (2–17).

The often-used technique of estimating dy/dx directly from a graph of y against x is probably the least accurate of the differentiation procedures, because of the difficulty in accurately locating a tangent to the graph at a chosen point. In this respect, some assistance is provided by first constructing the normal to the curve, following which the tangent may be drawn perpendicular to the normal. Location of the normal is facilitated by the use of a front-surfaced mirror placed perpendicular to the paper and revolved with its front edge remaining on the point at which the normal is to be constructed. The mirror is in the normal position when the graph and its reflection appear to form one smooth, continuous curve. Most accurate results are obtained when the tangent or the normal is close to an angle of 45°.

Of the methods available for numerical integration, the *trapezoidal method* represents the simplest, and, for scattered data, is frequently the most reliable. The method assumes that the area under a curve may be segmented into a series of adjoining trapezia, the area of each trapezium being given by the product of the width, Δx, and the average height, $(y_n + y_{n+1})/2$. For constant x intervals, the method yields the following relationship:

$$\int_{x_0}^{x_n} y\, dx = \Delta x(y_0/2 + y_1 + y_2 + \cdots + y_{n-1} + y_n/2) \tag{2–19}$$

The trapezoidal method assumes the experimental points which define a curve to be joined by a sequence of straight lines. In principle, the arc of a curve should be more accurately represented by a higher degree polynomial, such as Eqn (2–5). The *Simpson one-third method*, which is based upon integration of the first three terms of the Gregory–Newton equation (2–13), assumes quadratic arcs for successive groups of three points. The method is applicable over an odd number of data points separated by equal x intervals, and is represented by Eqn (2–20) (in which n is even).

$$\int_{x_0}^{x_n} y\, dx = (\Delta x/3)[(y_0 + y_n) + 4(y_1 + y_3 + y_5 + \cdots + y_{n-1})$$
$$+ 2(y_2 + y_4 + y_6 + \cdots + y_{n-2})] \tag{2–20}$$

Further extensions of Simpson's method are available in which the data are represented by a succession of adjoining cubic or higher degree arcs. The use of these methods is only justified for very smooth data. As may be observed from Figs. 2–2 and 2–3, Simpson's method has little or no advantage over the simple trapezoidal method in the case of scattered data.

2–4   Equation fitting

The choice of an equation to represent data may be quite empirical or it may be based upon a model. The latter represents the more useful basis of choice, but it should be appreciated that a given set of data is expressible in terms of a variety of equation types, some fitting the data more closely than others, some being more complicated in form than others. The ideal equation is one which represents the data closely, has a small number of arbitrary constants, and is evolved from a theoretical model (that is, rational rather than empirical). Usually, however, a compromise needs to be made among these desirable features. Models used in the field of chemical kinetics are discussed in Chapter 3. The selection of a suitable empirical equation is based largely upon experience and the experimenter's knowledge of graph shapes. In the case of polynomial equations, the tabular suitability test, mentioned earlier in this chapter, provides a valuable indicator for selection.

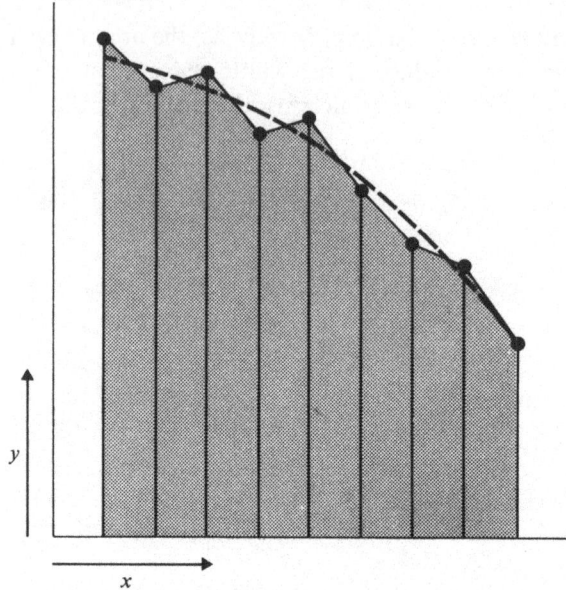

Fig. 2–2 The trapezoidal method applied to scattered data. The dashed line represents a possible smooth curve through the data points.

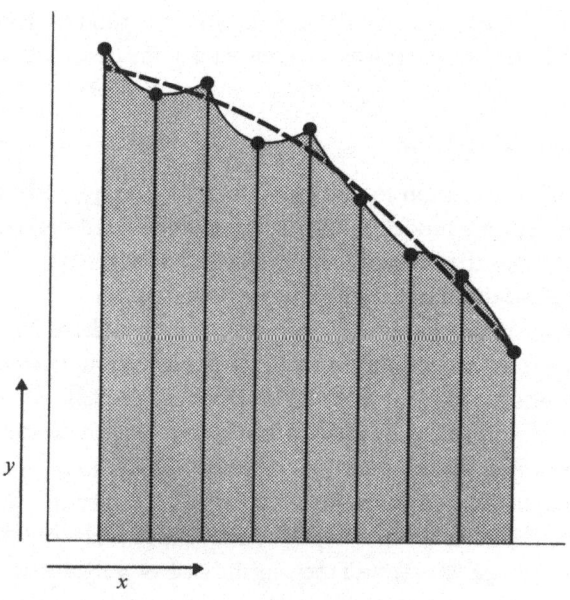

Fig. 2–3 Simpson's one-third method applied to scattered data.

Given a set of data and a form of equation, the question arises as to the method by which the arbitrary constants may best be evaluated. When the value of the constants only is desired, the *method of selected points* or the *method of averages* is usually adequate; when some estimate of error or confidence limits of the arbitrary constants is required, the more cumbersome *method of least squares* is usually preferred. However, a more meaningful judgement of the accuracy of estimate of the constants is likely to emerge from comparisons among a number of experiments rather than from data obtained from single experiments. This point of argument is particularly relevant to the estimate of a rate coefficient, k, in a chemical kinetic study—there is little purpose in attempting an estimate of error in the value of k, as judged from the *time/concentration* data from one particular kinetic run, when this value is likely to be more greatly affected by such changes in conditions from run to run as fluctuations in the temperature or the presence of trace contaminants.

In applying the method of selected points, a number of data values are chosen equal to the number of arbitrary constants in the selected equation. The choice should be such that values are well separated and are representative of the shape of the curve over its length. These values are then substituted in the general equation to give a series of simultaneous equations with the arbitrary constants as unknowns. Solution of the simultaneous equations yields estimates for the arbitrary constants. The overall equation so obtained should be checked by substituting, in turn, each x value in the original data, and comparing the y values so calculated with those experimentally observed. On the basis of these comparisons, slight adjustments may be made to the arbitrary constants in order to obtain an improved fit. The method works best for smooth data, and, in this respect, annoying influence of scatter upon the selection of points may be reduced by preliminary smoothing in the region from which the points are to be selected. An illustration of this method is given in Example 2–7, at the end of this chapter.

The method of averages is similar to the method of selected points, but the selection is made by averaging x and y coordinates within each chosen group of data points. The method is particularly appropriate to straight-line data for which it yields results comparable with those obtained by the least-squares procedure. The first step in the procedure for fitting to the straight-line equation, $y = a + bx$, is to divide the total data into two groups containing m and n points respectively: $(x_1, y_1; x_2, y_2; \ldots; x_m, y_m)$ and $(x'_1, y'_1; x'_2, y'_2; \ldots; x'_n, y'_n)$. An estimate of b is then given by Eqn (2–21):

$$b = \frac{(y_1 + y_2 + \cdots + y_m)/m - (y'_1 + y'_2 + \cdots + y'_n)/n}{(x_1 + x_2 + \cdots + x_m)/m - (x'_1 + x'_2 + \cdots + x'_n)/n} \tag{2–21}$$

The corresponding estimate of a is obtained by substituting the estimated value of b, the average of all x values ($\bar{x}$), and of y values ($\bar{y}$) in the straight-line equation,

$$\bar{y} = a + b\bar{x} \qquad (2\text{-}22)$$

When the original data are distributed in a nearly even pattern, m and n should be taken as nearly equal. For an even number of such data points it is convenient to take m as equal to n, for then Eqn (2–21) simplifies to the form

$$b = \frac{(y_1 + y_2 + \cdots + y_m) - (y'_1 + y'_2 + \cdots + y'_m)}{(x_1 + x_2 + \cdots + x_m) - (x'_1 + x'_2 + \cdots + x'_m)} \qquad (2\text{-}23)$$

For an odd number of data points, little error is normally involved in excluding the median value for the estimate of b (allowing Eqn (2–23) to be used), and including it for the estimate of a. The principle of the method of averages is illustrated in pictorial form by Fig. 2–4, and its application to a numerical case is given in Example 2–8 at the end of this chapter.

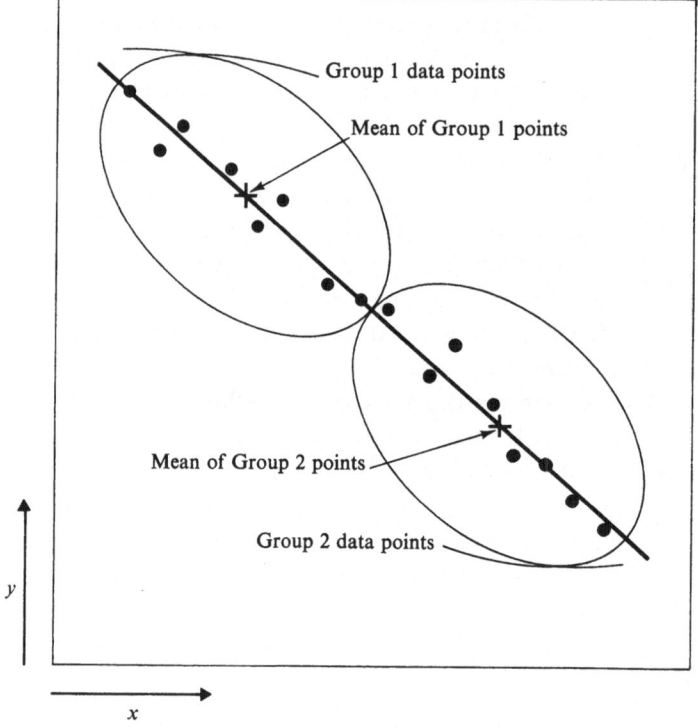

Fig. 2–4   Fitting a straight line by the method of averages.

The method of least squares is based upon the concepts of the gaussian distribution outlined in Chapter 1. For univariate data, the mean corresponds to that value for which the sum of the squares of the deviations is a minimum, and, accordingly, for associated (bivariate) data, the least-squares line corresponds to that line for which the sum of the squares of the deviations (of the recorded data from the line) is also a minimum. It is clear that the probability concepts connected with the gaussian distribution are also applicable to this case, but the fitting of the line to data depends upon the way in which the deviations are measured. The usual assumption is to take the x values as being precise and to tie the deviations only to the y values. The least-squares line so fitted is different from that constructed on the assumption that both x and y values are subject to error.

If, for each of a series of known x values, a number of estimates of the corresponding y values are available, these may be arranged in a series of frequency distributions as illustrated in perspective in Fig. 2–5.

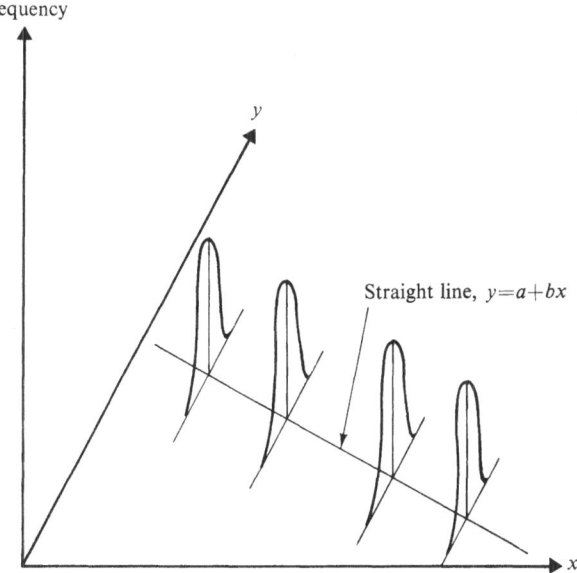

Fig. 2–5   Frequency distributions of estimates of y values for known values of x (shown in perspective).

In this figure it is assumed that the data may be represented by a straight line, $y = a + bx$. In attempting to find the position of this line, the following simplifying assumptions may be made:

(a) that only y values are subject to error;
(b) that each distribution of the y value for a given x corresponds to a gaussian type;
(c) that each distribution has the same value of σ;
(d) that the means of the distributions lie on the one straight line.

In summary form, it is assumed that all y deviations from the line belong to a common gaussian distribution. Under these conditions, the best estimate of the straight line is obtained when the sum of the squares of the y deviations from the line is a minimum. This 'best estimate' is expressed in terms of numerical values for a and b in the straight-line equation, $y = a + bx$, and, following the principles outlined in Chapter 1, confidence limits may be associated with these values. These measures of reliability of the a and b values are, however, not independent of each other, but represent a joint confidence measure for slope and intercept on the y axis. Calculation of these limits is made with the aid of t values as listed in Table 1–3, and it should be remembered that, for straight-line fitting, the number of degrees of freedom is two less than the number of data points, as the minimum number of points required to define a straight line is two. The nature of the dispersion of points about the line does not become evident until more than this minimum number of points are available.

An illustration of least-squares fitting according to the principles just outlined is given in Example 2–9 at the end of this chapter. In this example, rate coefficients for a chemical reaction studied over a range of temperatures are fitted to the Arrhenius equation in the form,

$$\log k = \log A - E/2 \cdot 303 R T$$

Application of the *simplified least-squares method* just outlined does not always lead to estimates of the constants in a straight-line equation which are truer than those obtained by use of the method of averages (see, for example, discussion in the next section of this chapter). However, its relationship to the gaussian distribution justifies its use as a standard method of coding in terms of measures of central tendency and reliability.

A number of extensions to this basic least-squares model is available. For example, differing values of σ may be assumed for the distributions shown in Fig. 2–5, and the y values may be weighted accordingly. Another extension, to which brief reference has already been made, is to assume that both x and y values are subject to error. Least-squares fitting may also be applied to curvilinear equations, such as quadratics and higher degree polynomials of the form given in Eqn 2–5. Interested readers will find further information in the texts to which reference has been made at the end of the chapter.

2–5   Propagation of error

In many situations, a required quantity is not available by direct measurement, but may be estimated from an equation involving other measurable quantities. Typical examples in the field of chemical kinetics are the estimate of the rate coefficient of a chemical reaction from time and concentration measurements, and the estimate of activation energy from rate and temperature measurements.

If y represents the required quantity, and x, z, etc. the measurable quantities (assumed independent), then in the form,

$$y = f(x, z, \text{etc.}) \tag{2-24}$$

and

$$dy = (\partial y/\partial x)\, dx + (\partial y/\partial z)\, dz + \cdots \tag{2-25}$$

Equation (2–25), sometimes called the *partial differential theorem* may be written in the following approximate form for small finite changes in y, x, z, etc.

$$\delta y = (\partial y/\partial x)\, \delta x + (\partial y/\partial z)\, \delta z + \cdots \tag{2-26}$$

This form of the equation is useful for estimating a specific change in y caused by specific changes in x, z, etc. the sign of the changes also being specified. If δx, δz, etc. correspond to possible errors in x, z, etc. with sign unspecified, the maximum possible error in y can be obtained by assuming that the signs of δx, δz, etc. cooperate in the most detrimental manner. This, however, presents a pessimistic view, and a better estimate of the likely error in y is given by Eqn (2–27), in which the symbol ε represents a statistical measurement of error,* such as the standard error.

$$\varepsilon_y = \sqrt{[(\partial y/\partial x)^2\, \varepsilon_{2x} + (\partial y/\partial z)^2\, \varepsilon_{2z} + \cdots]} \tag{2-27}$$

It is instructive to apply the error analysis, just outlined, to the particular case of a chemical decomposition proceeding by a first-order kinetic law. Consider the reaction to be proceeding in solution, and the rate coefficient, k, to be estimated from the equation,

$$k = (1/t) \ln (C_0/C) \tag{2-28}$$

The concentration, C, of the decomposing substance is taken as being measured by titration of a standard-sized sample after a measured period of time, t, the initial concentration being represented by C_0. The measurements of t, C_0, and C are subject to error, and application of Eqn (2–26) gives,

---

*For readings from an instrument, the standard error may be approximated to two-thirds of the estimated maximum error of reading.

$$\delta k = (\partial k/\partial t)\, \delta t + (\partial k/\partial C_0)\, \delta C_0 + (\partial k/\partial C)\, \delta C \qquad (2\text{–}29)$$

$$= (1/t)\left[-(\delta t/t)\ln (C_0/C) + (\delta C_0/C_0) - (\delta C/C) \right] \qquad (2\text{–}30)$$

The maximum positive error in k therefore occurs when the errors in t, C_0, and C are negative, positive, and negative, respectively

For the purpose of a numerical calculation with Eqn (2–30), assume a half-life of 1000 s, with $\delta t = \pm 1$ s; further, assume that C_0 is determined by a titration of 20·0 ml of reagent and that the possible error in titrating is $\pm 0\cdot 1$ ml. Substitution in Eqn (2–30) (with appropriate sign adjustment) gives the maximum error in k at the half-life ($C_0/C = 2$) as,

$$\delta k_{max} = \pm 1\cdot 6 \times 10^{-5}\ \text{s}^{-1}$$

for a value of k equal to $69\cdot 3 \times 10^{-5}\ \text{s}^{-1}$. Examination of Table 2–5, which lists maximum errors for other values of C_0/C, shows that the error in estimating k varies with the value of C_0/C, and is at a minimum near $C_0/C = 3\cdot 5$, that is at about 70% decomposition of the substance.

Table 2–5 **Illustrating propagation of error for first-order decomposition**

C_0/C	Decomposition (%)	t (s)	$\delta t/t$ ($\times 10^4$)	$\delta C/C$ ($\times 10^3$)	δk (s^{-1}) ($\times 10^5$)
1·0	0	0	—	—	—
1·3	23	379	26·4	6·5	3·21
1·6	38	678	14·8	8·0	2·02
2·0	50	1000	10·0	10·0	1·57
2·5	60	1323	7·6	12·5	1·38
3·0	67	1586	6·3	15·0	1·30
3·5	71	1810	5·5	17·5	1·28
4·0	75	2000	5·0	20·0	1·29
5·0	80	2325	4·3	25·0	1·32
7·0	86	2810	3·6	35·0	1·45
10·0	90	3325	1·1	50·0	1·71

It is assumed that the reaction has a half-life of 1000 s, $\delta t = \pm 1$ s, C is determined by titration with a possible titration error of $\pm 0\cdot 1$ ml, and that the original titre = 20·0 ml for C_0.

The rate coefficient for a first-order reaction is commonly estimated from the slope of the straight-line graph of $\log (C_0/C)$ plotted against t. If the line is fitted by the simplified least-squares method, it is assumed that the values for t are known with certainty and that the recorded values for $\log (C_0/C)$ are subject to the same likelihood of error over the range of data (see page 30). The latter assumption is inconsistent with the results of the error analysis just performed, and, if the magnitude only of the rate coefficient is required, fitting of the straight line by the method of averages is no less satisfactory than by the simplified least-squares method. A more correct procedure for fitting would involve the assumptions that the likelihood of error is constant for the recorded values of t, and is variable for the values of $\log(C_0/C)$, this variation being assessed in terms of an error analysis similar to the one previously carried out. Such a fitting procedure would of course be considerably more complicated than either of the two just discussed. As pointed out earlier in this chapter, however, the accuracy of the estimate of a rate coefficient is probably more meaningfully judged from the results of a number of experiments rather than from the data obtained from a single experiment.

The value of a particular numerical method should not be assessed in terms of its sophistication alone, but should be judged in as wide a context as possible. Experimenters in the field of chemical kinetics soon become aware (often all too painfully) that chemical systems are rarely, if ever, as pure as the models which are used to represent them!

Examples and problems

Example 2–1 The smoothing of data

Table 2–6 lists data for a gaseous chemical reaction which has been followed by the measurement of changes in the total pressure of gas in a closed reaction vessel. The third column in the table lists the first-order item differences in pressure, Δp, and the plot of these against time, as shown in Fig. 2–6, indicates that the original data are not smooth.

From the curve drawn through the points shown in Fig. 2–6, smooth values of the item differences, $\Delta' p$, are estimated, and these are listed in the fourth column of Table 2–6. Assuming the first p value as being correct, the smoothed differences are added in a cumulative manner to yield the trial set of smoothed p values listed in the fifth column of the table. The mean value of these shows a significant positive deviation from the mean of the original data, and the trial values are accordingly adjusted to yield the final set of smoothed values, p', shown in the last column of the table.

For the smoothing of still more accurate data, the second-order item differences may be treated in a similar fashion.

Example 2–2

Assuming the values for C listed in Table 2–3 corresponding to $t = 20, 40, 60, 80, 100$, an estimate of C for $t = 50$ may be made by the linear method as follows:

Linear interpolation (Eqn (2–12))
Let $x_n = 50$. Take $x_1 = 40$, $x_2 = 60$ (t values). Let $y_n = C_{(t=50)}$. Take $y_1 = 36\cdot4$, $y_2 = 31\cdot9$ (C values).

$$y_n = y_1 + (y_2 - y_1)(x_n - x_1)/(x_2 - x_1) \quad \text{Eqn (2–12)}$$

$$= 36\cdot4 + (31\cdot9 - 36\cdot4)(50 - 40)/(60 - 40)$$

$$= 36\cdot4 - 2\cdot2_5$$

$$= 34\cdot1_5$$

Table 2–6   Smoothing by the item difference method

(See also Example 2–1 and Fig. 2–6)

t = reaction time in minutes; p = total pressure (cm Hg); $\Delta'p$ = smoothed first-order item differences in p; p'_{trial} = smoothed values of p, first trial; p' = final smoothed values of p.

t	p	Δp	$\Delta'p$	p'_{trial}	p'
4	21·07			21·07	21·06
		0·92	0·94		
8	21·99			22·01	22·00
		0·86	0·88		
12	22·85			22·89	22·88
		0·85	0·82		
16	23·70			23·71	23·70
		0·76	0·77		
20	24·46			24·48	24·47
		0·74	0·73		
24	25·20			25·21	25·20
		0·67	0·69		
28	25·87			25·90	25·89
		0·66	0·65		
32	26·53			26·55	26·54
		0·64	0·62		
36	27·17			27·17	27·16
		0·58	0·59		
40	27·75			27·76	27·75
		0·58	0·56		
44	28·33			28·32	28·31
		0·53	0·54		
48	28·86			28·86	28·85

Mean = 25·31$_4$ Mean = 25·32$_8$

Linear extrapolation (Eqn (2–12))

Take $x_1 = 60$, $x_2 = 80$ and $y_1 = 31·9$, $y_2 = 28·3$. Applying the formula as before,

$$y_n = 33·7$$

The mean of the interpolated and extrapolated values is $C_{(t=50)} = 33·9$ (to the first decimal place). Table 2–3 lists the correct value for C at $t = 50$ as 34·0.

Example 2–3

Using the same data as in Example 2–2, an estimate of C for $t = 50$ (Table 2–3) may be made by the Gregory–Newton method, the method being applied for equally spaced values of the independent variable, t.

Gregory–Newton interpolation (Eqn (2–13))

Let $x_n = 50$. Take $x_0 = 40$; $x_1 = 60$; $x_2 = 80$; $x_3 = 100$.

$$n = (x_n - x_0)/\Delta x = (50 - 40)/20 = 0·5$$

Take $y_0 = 36·4$; $y_1 = 31·9$; $y_2 = 28·3$; $y_3 = 25·5$ (Table 2–3).

$$\Delta y_0 = y_1 - y_0 = -4·5; \qquad \Delta y_1 = y_2 - y_1 = -3·6$$

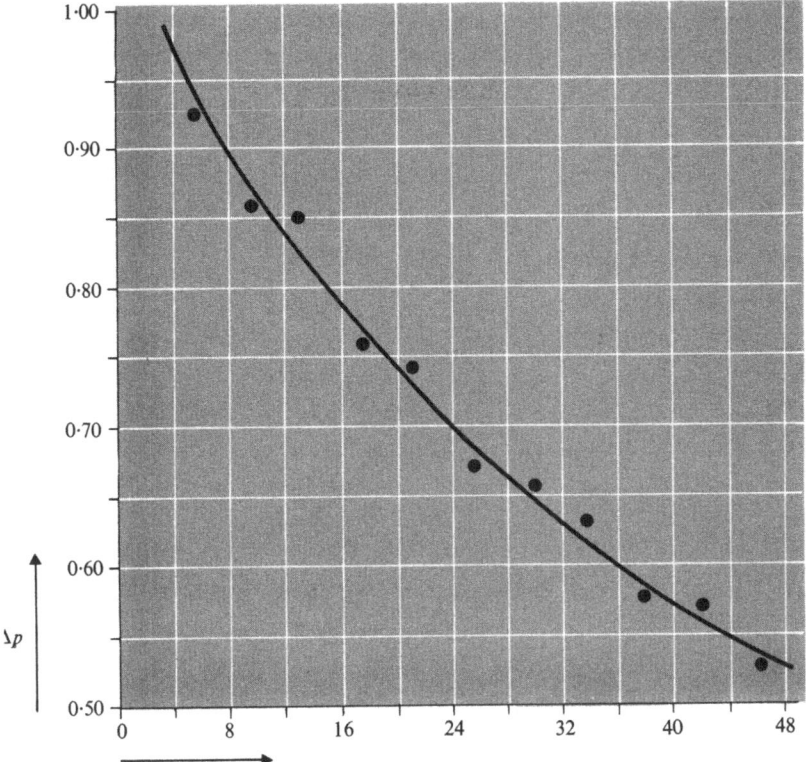

Fig. 2–6   Smoothing first-order item differences (see Table 2–6 for the data).

$$\Delta y_2 = y_3 - y_2 = -2\cdot8$$

$$\Delta^2 y_0 = \Delta y_1 - \Delta y_0 = 0\cdot9; \qquad \Delta^2 y_1 = \Delta y_2 - \Delta y_1 = 0\cdot8$$

$$\Delta^3 y_0 = \Delta^2 y_1 - \Delta^2 y_0 = -0\cdot1$$

$$y_n = y_0 + n\,\Delta y_0 + n(n-1)\,\Delta^2 y_0/2 + n(n-1)(n-2)\,\Delta^3 y_0/6 \quad \text{Eqn (2–13)}$$

$$= 36\cdot4 + 0\cdot5(-4\cdot5) + 0\cdot5(-0\cdot5)(0\cdot9)/2$$

$$+ 0\cdot5(-0\cdot5)(-1\cdot5)(-0\cdot1)/6$$

$$= 36\cdot4 - 2\cdot2_{50} - 0\cdot1_{12} - 0\cdot0_{06} = 34\cdot0_3$$

Therefore, by Gregory–Newton interpolation, $C = 34\cdot0$ for $t = 50$. (This is the same as the value listed in Table 2–3.)

Example 2–4

Assuming the values for C listed in Table 2–3 for $t = 10, 30, 40, 60$ an estimate of C for $t = 0$ may be made by the *Gregory–Newton divided difference method*. This method is appropriate, as the C values are not equally spaced.

Gregory–Newton divided difference formula (Eqn (2–14))
Let $x_n = 0$. Take $x_0 = 10$; $x_1 = 30$; $x_2 = 40$; $x_3 = 60$ and $y_0 = 46\cdot4$; $y_1 = 39\cdot2$; $y_2 = 36\cdot4$; $y_3 = 31\cdot9$ (Table 2–3).

Divided differences are given by Eqns (2–6) to (2–11):

$$\Delta_d y_0 = (y_1 - y_0)/(x_1 - x_0) = (-7\cdot2)/20 = -0\cdot360$$

$$\Delta_d y_1 = (y_2 - y_1)/(x_2 - x_1) = (-2\cdot8)/10 = -0\cdot280$$

$$\Delta_d y_2 = (y_3 - y_2)/(x_3 - x_2) = (-4\cdot5)/20 = -0\cdot225$$

$$\Delta_d^2 y_0 = (\Delta_d y_1 - \Delta_d y_0)/(x_2 - x_0) = 0\cdot08/30 = 2\cdot667 \times 10^{-3}$$

$$\Delta_d^2 y_1 = (\Delta_d y_2 - \Delta_d y_1)/(x_3 - x_1) = 0\cdot055/30 = 1\cdot833 \times 10^{-3}$$

$$\Delta_d^3 y_0 = (\Delta_d^2 y_1 - \Delta_d^2 y_0)/(x_3 - x_0) = -0\cdot834 \times 10^{-3}/50$$

$$= -1\cdot67 \times 10^{-5}.$$

$$y_n = y_0 + (x_n - x_0)\,\Delta_d y_0 + (x_n - x_0)(x_n - x_1)\,\Delta_d^2 y_0$$

$$\qquad + (x_n - x_0)(x_n - x_1)(x_n - x_2)\,\Delta_d^3 y_0 \quad \text{Eqn (2–14)}$$

$$= 46\cdot4 + (-10)(-0\cdot36) + (-10)(-30)(2\cdot667 \times 10^{-3})$$

$$\qquad + (-10)(-30)(-40)(-1\cdot67 \times 10^{-5})$$

$$= 46\cdot4 + 3\cdot6_{00} + 0\cdot8_{00} + 0\cdot2_{00}$$

$$= 51\cdot0_0.$$

Therefore, by Gregory–Newton extrapolation, $C = 51\cdot0$ for $t = 0$. (This is the same as the value listed in Table 2–3.)

Example 2–5

Using the same data as in Example 2–4, an estimate of C for $t = 0$ (Table 2–3) may be made by the Lagrange method as follows:

Lagrange extrapolation Eqn (2–15):
Let $x_n = 0$. Take $x_1 = 10$; $x_2 = 30$; $x_3 = 40$; $x_4 = 60$ and $y_1 = 46\cdot4$; $y_2 = 39\cdot2$; $y_3 = 36\cdot4$; $y_4 = 31\cdot9$ (Table 2–3).

$$y_n = y_1\left(\frac{x_n - x_2}{x_1 - x_2} \times \frac{x_n - x_3}{x_1 - x_3} \times \frac{x_n - x_4}{x_1 - x_4}\right)$$

$$+ y_2\left(\frac{x_n - x_1}{x_2 - x_1} \times \frac{x_n - x_3}{x_2 - x_3} \times \frac{x_n - x_4}{x_2 - x_4}\right)$$

$$+ y_3\left(\frac{x_n - x_1}{x_3 - x_1} \times \frac{x_n - x_2}{x_3 - x_2} \times \frac{x_n - x_4}{x_3 - x_4}\right)$$

$$+ y_4\left(\frac{x_n - x_1}{x_4 - x_1} \times \frac{x_n - x_2}{x_4 - x_2} \times \frac{x_n - x_3}{x_4 - x_3}\right) \quad \text{Eqn (2–15)}$$

$$= 46\cdot4\left(\frac{-30}{-20} \times \frac{-40}{-30} \times \frac{-60}{-50}\right)$$

$$+ 39\cdot2\left(\frac{-10}{20} \times \frac{-40}{-10} \times \frac{-60}{-30}\right)$$

$$+ 36\cdot4\left(\frac{-10}{30} \times \frac{-30}{10} \times \frac{-60}{-20}\right)$$

$$+ 31\cdot9\left(\frac{-10}{50} \times \frac{-30}{30} \times \frac{-40}{20}\right)$$

$$= (46\cdot4 \times 12/5) - (39\cdot2 \times 4) + (36\cdot4 \times 3) - (31\cdot9 \times 2/5)$$

$$= 111 \cdot 3_6 - 156 \cdot 8 + 109 \cdot 2 - 12 \cdot 7_6$$

$$= 51 \cdot 0_0$$

By Lagrange extrapolation, therefore, $C = 51 \cdot 0$ for $t = 0$, in agreement with the value listed in Table 2–3.

Example 2–6

From the data in Table 2–3 (assuming the value for C at $t = 0$ is not known) an estimate of $-dC/dt$ at time, $t = 0$, may be made as follows.

Differentiated Gregory–Newton method (Eqn (2–16))
Let $x_n = 0$. Take $x_0 = 10$ (t values). With $\Delta x = 10$, $n = (x_n - x_0)/\Delta x = -1$; $y_0 = 46 \cdot 4$; $\Delta y_0 = -3 \cdot 9$; $\Delta^2 y_0 = +0 \cdot 6$; $\Delta^3 y_0 = -0 \cdot 1$ (reading from the difference table for C).

$$dy/dx = (1/\Delta x)[\Delta y_0 + (2n-1)\,\Delta^2 y_0/2 + (3n^2 - 6n + 2)\,\Delta^3 y_0/6 + \cdots]$$
$$\text{Eqn (2–16)}$$

$$= (1/10)[-3 \cdot 9 + (-3)(0 \cdot 6)/2 + (11)(-0 \cdot 1)/6 + \cdots]$$

$$= (1/10)[-3 \cdot 9 - 0 \cdot 9 - 0 \cdot 2]$$

$$-dy/dx = (1/10)(5 \cdot 0) = 0 \cdot 50$$

$$-dC/dt_{(\text{for } t = 0)} = 0 \cdot 50 \text{ millimole } l^{-1}\,s^{-1}$$

$$= 5 \cdot 0 \times 10^{-4} \text{ mole } l^{-1}\,s^{-1}$$

Example 2–7

The following illustrates the *method of selected points* for fitting the data in Table 2–2 to an empirical equation of the form $y = a + bx + cx^2$: choose reference points at $x = 1 \cdot 1$; $1 \cdot 4$; $1 \cdot 7$ and $y = 3 \cdot 44$; $4 \cdot 52$; $5 \cdot 24$. Substitution of these values in the required general equation leads to the following three simultaneous equations:

$$3 \cdot 44 = a + b(1 \cdot 1) + c(1 \cdot 1)^2 \quad (1)$$

$$4 \cdot 52 = a + b(1 \cdot 4) + c(1 \cdot 4)^2 \quad (2)$$

$$5 \cdot 24 = a + b(1 \cdot 7) + c(1 \cdot 7)^2 \quad (3)$$

Solution of these three equations yields the following estimates of the parameters in the empirical equation,

$$a = -3 \cdot 60$$

$$b = 8 \cdot 6$$

$$c = -2 \cdot 0$$

The required equation (first estimate) is therefore,

$$y = -3 \cdot 60 + 8 \cdot 6x - 2 \cdot 0x^2$$

This equation is tested by substituting, in turn, each of the x values listed in the original table (Table 2–2), and comparing the y values, so calculated, with those listed in the table. The differences between these two sets of y values are given in Table 2–7.

It is evident that there is good agreement between the data values and those calculated from the empirical equation, and therefore, no adjustment of the parameters in the equation is required in this instance. If the residuals, listed in the last column of Table 2–7, were to show an average bias away from zero or a definite trend, then adjustment of the parameters in the empirical equation would be necessary. In the

Table 2–7   Checking an empirical equation against the original data

(See Example 2–7 and Table 2–2.)

x	y (data)	y (equation)	y (data) $- y$ (equation)
1·0	3·01	3·00	+0·01
1·1	3·44	3·44	0·00
1·2	3·84	3·84	0·00
1·3	4·19	4·20	−0·01
1·4	4·52	4·52	0·00
1·5	4·80	4·80	0·00
1·6	5·05	5·04	+0·01
1·7	5·24	5·24	0·00
1·8	5·40	5·40	0·00

case of a polynomial, this may be readily effected by fitting the residuals also to a polynomial of the form

$$y \text{ (residual)} = y \text{ (data)} - y \text{ (equation)} = a^1 + b^1 x + c^1 x^2 + \cdots \tag{2-31}$$

and adding the values of a^1, b^1, c^1, etc. to those of a, b, c, etc. obtained in the first estimate of the empirical equation to yield a second estimate of the equation. Similar adjustment procedures may be adopted for other forms of empirical equations.

Example 2–8

Values of the first-order rate coefficients for the decomposition, over a range of temperatures, of cyclopentyl chloride to cyclopentene and hydrogen chloride are listed in Table 2–8.

Estimates of the parameters, A and E, in the Arrhenius equation,
   $\log k = \log A - E/2 \cdot 303 R T$,

Table 2–8   Variation of a rate coefficient with temperature

(Thermal decomposition of cyclopentyl chloride)
T = temperature in degrees Kelvin; k = first-order rate coefficient in s^{-1}.

Experiment	T	$10^5 \times k$	$10^3/T$	$5 + \log_{10} k$
1	582·2	2·7$_{27}$	1·7176	0·4357
2	583·7	2·9$_{14}$	1·7132	0·4645
3	583·8	3·0$_{41}$	1·7129	0·4830
4	599·6	8·7$_{10}$	1·6678	0·9400
5	599·9	8·1$_{06}$	1·6669	0·9088
6	615·6	23·$_{22}$	1·6244	1·3659
7	615·8	23·$_{53}$	1·6239	1·3717
8	632·5	68·$_{19}$	1·5810	1·8337
9	632·5	66·$_{51}$	1·5810	1·8229
10	632·6	70·$_{96}$	1·5808	1·8510

may be made by the method of averages as follows: in the straight-line equation, $y = a + bx$, let

$$y = 5 + \log k$$

$$a = 5 + \log A$$

$$b = -10^{-3}E/2 \cdot 303R$$

$$x = 10^3/T$$

(The introduction of the 5, 10^3, and 10^{-3} terms leads to figures of a convenient magnitude for computation.) The data in Table 2–8 are divided into two groups, namely Expt 1 to 5 and Expt 6 to 10.

$$b = \frac{(y_1 + y_2 + y_3 + y_4 + y_5) - (y_6 + y_7 + y_8 + y_9 + y_{10})}{(x_1 + x_2 + x_3 + x_4 + x_5) - (x_6 + x_7 + x_8 + x_9 + x_{10})}$$

(Eqn (2–23), for a total of ten data points.)

y values

(1)	0·4357	(6)	1·3659
(2)	0·4645	(7)	1·3717
(3)	0·4830	(8)	1·8337
(4)	0·9400	(9)	1·8229
(5)	0·9088	(10)	1·8510

Totals 3·2320 + 8·2452 = 11·4772

Average $\bar{y} = 11\cdot4772/10 = 1\cdot14772$

Difference $\sum(y_1 \text{ to } y_5) - \sum(y_6 \text{ to } y_{10}) = -5\cdot0132$

x values

(1)	1·7176	(6)	1·6244
(2)	1·7132	(7)	1·6239
(3)	1·7129	(8)	1·5810
(4)	1·6678	(9)	1·5810
(5)	1·6669	(10)	1·5808

Totals 8·4784 + 7·9911 = 16·4695

Average $\bar{x} = 16\cdot4695/10 = 1\cdot64695$

Difference $\sum(x_1 \text{ to } x_5) - \sum(x_6 \text{ to } x_{10}) = 0\cdot4873$

$$b = -5\cdot0132/0\cdot4873$$

$$E = 10^3 \times 5\cdot0132 \times 2\cdot303 \times 1\cdot9871/0\cdot4873 \text{ cal mole}^{-1}$$

$$= 47\,070 \text{ cal mole}^{-1}$$

Using $\bar{y} = a + b\bar{x}$ (Eqn (2–22)),

$$1\cdot14772 = a - 5\cdot0132 \times 1\cdot64695/0\cdot4873$$

$$a = 18\cdot08$$

$$\log A = 13\cdot08$$

$$A = 1\cdot20 \times 10^{13} \text{ s}^{-1}$$

The Arrhenius parameters by the method of averages are therefore $A = 1\cdot20 \times 10^{13} \text{ s}^{-1}$ and $E = 47\,070 \text{ cal mole}^{-1}$.

Example 2–9

The simplified least-squares method may be applied to the data in Table 2–8 and Example 2–8 to yield estimates of the parameters A and E in the Arrhenius equation, together with their 90% confidence limits. In this instance, the temperatures are assumed correct, and all errors are assumed to reside in the recorded values of k, each of which is given equal weighting. The basic assumptions of the simplified least-squares method are given in page 30 of this chapter. (A more sophisticated approach to the application of least-squares principles to data of this type has been discussed by W. R. McBride and D. S. Villars in *Analyt. Chem.*, 1954, **26**, 901.)

As in Example 2–8, let $y = 5 + \log k$ and $x = 10^3/T$. In computing the least-squares line, values of y, x, y^2, x^2, and xy are listed as shown in Table 2–9. Values of $(x+y)^2$ are also useful for checking purposes. The check is made for

$$\sum x^2 + \sum y^2 + 2 \sum xy = \sum (x+y)^2$$

If these values, as computed in the table, do not agree, then individual rows in the table should be checked for

$$x^2 + y^2 + 2xy = (x+y)^2$$

Table 2–9   Least-squares computation of Arrhenius parameters

(See data in Table 2–8; further details of the computation are given in Example 2–9)

Expt	y	x	y^2	x^2	xy	$(x+y)^2$
1	0·4357	1·7176	0·1898345	2·9501498	0·7483583	4·6367009
2	0·4645	1·7132	0·2157602	2·9350542	0·7957814	4·7423773
3	0·4830	1·7129	0·2332890	2·9340264	0·8273307	4·8219768
4	0·9400	1·6678	0·8836000	2·7815568	1·5677320	6·8006208
5	0·9088	1·6669	0·8259174	2·7785556	1·5148787	6·6342305
6	1·3659	1·6244	1·8656828	2·6386754	2·2187680	8·9418941
7	1·3717	1·6239	1·8815609	2·6370512	2·2275036	8·9736194
8	1·8337	1·5810	3·3624557	2·4995610	2·8990797	11·6601761
9	1·8229	1·5810	3·3229644	2·4995610	2·8820049	11·5865352
10	1·8510	1·5808	3·4262010	2·4989286	2·9260608	11·7772512
Totals	11·4772	16·4695	16·2072659	27·1531200	18·6074981	80·5753823

Mean values $\bar{y} = 1·14772$ $\bar{x} = 1·64695$

Squared means $(\bar{y})^2 = 1·31726120$ $(\bar{x})^2 = 2·71244430$

Cross term $\bar{x}\bar{y} = 1·89023745$

Check on computation using $(x+y)^2$ term:

$$\sum x^2 + \sum y^2 + 2 \sum xy = 80·5753821$$

From the table,

$$\sum (x+y)^2 = 80·5753823$$

The check shows a satisfactory agreement.

As shown in Table 2–9, the mean values, $\bar{y}$ and $\bar{x}$, the squared means, $(\bar{y})^2$ and $(\bar{x})^2$, and the cross term, $\bar{x}\bar{y}$, are also computed. In place of the equation form, $y = a + bx$, it is simpler to convert to the form $Y = bX$. For fitting to the equation in this form, the following relationships are used:

$$\sum Y^2 = \sum y^2 - n(\bar{y})^2 \tag{2–32}$$

$$\sum X^2 = \sum x^2 - n(\bar{x})^2 \tag{2-33}$$

$$\sum XY = \sum xy - n\bar{x}\bar{y} \tag{2-34}$$

n = number of data points (in this example $n = 10$).
From the information in Table 2–9,

$$\sum Y^2 = 3{\cdot}0346539$$

$$\sum X^2 = 0{\cdot}0286770$$

$$\sum XY = -0{\cdot}2948764$$

The term, b, in the least-squares equation is given by

$$b = \sum XY/\sum X^2 \tag{2-35}$$

$$= -10{\cdot}2826794$$

The least-squares line passes through the point $\bar{x}$, $\bar{y}$, and the term, a, is therefore given by

$$a = \bar{y} - b\bar{x} \tag{2-36}$$

$$= 18{\cdot}0828$$

As in Example 2–8,

$$a = 5 + \log A \qquad \text{(Arrhenius } A\text{)}$$

$$b = -10^{-3}E/2{\cdot}303R \qquad \text{(Arrhenius } E\text{)}$$

Therefore

$$\log A = 13{\cdot}0828$$

$$A = 1{\cdot}210 \times 10^{13} \text{ s}^{-1}$$

$$E = 47\,046 \text{ cal mole}^{-1}$$

Confidence limits of A *and* E
An estimate of the standard error of the term, b, is given by Eqn (2–37):

$$\text{s.e. } (b) = [(\sum Y^2 - b \sum XY)/(n-2) \sum X^2]^{1/2} \tag{2-37}$$

$$= 0{\cdot}1051$$

The term, $n-2$, in Eqn (2–37) represents the number of degrees of freedom which, in the case of a straight-line fit, is two less than the total number of data points.

$$\text{Confidence limits of } b = \pm t \times \text{s.e. } (b) \tag{2-38}$$

(Compare with Eqn (1–10).)
From Table 1–3, $t = 1{\cdot}86$, for 90% limits and eight degrees of freedom. Therefore

90% confidence limits of $b = \pm 0{\cdot}195$

90% confidence limits of $E = \pm 890$ cal mole^{-1}

An estimate of the standard error of the term, a, is given by Eqn (2–39):

$$\text{s.e. } (a) = \left[(\sum Y^2 - b \sum XY) \frac{1/n + \bar{x}^2/\sum X^2}{n-2} \right]^{1/2} \tag{2-39}$$

$$= 0{\cdot}1732$$

$$\text{Confidence limits of } a = \pm t \times \text{s.e. } (a) \tag{2-40}$$

90% confidence limits of $a = \pm 0{\cdot}322$

90% confidence limits of $\log A = \pm 0{\cdot}322$

The estimated values of E and $\log A$ together with the corresponding 90% confidence limits are therefore:

$$E = (47050 \pm 890)\, \text{cal mole}^{-1}$$

$$\log A = 13\cdot08 \pm 0\cdot32$$

that is

$$A = 1\cdot21 \times 10^{13}\, \text{s}^{-1}$$

with the upper 90% confidence limiting value at $2\cdot54 \times 10^{13}\, \text{s}^{-1}$, and the lower 90% confidence limiting value at $5\cdot77 \times 10^{12}\, \text{s}^{-1}$. Note that although the confidence limits are equally spaced on either side of the $\log A$ value, they are unequally spaced about the corresponding A value. This is because the least-squares line has been calculated with the deviations of the $\log k$ values from the line approximating to a gaussian distribution (unskewed), and the corresponding deviations of the k values from those calculated by means of the fitted line therefore give a skewed distribution. Mention of this point was made in Chapter 1.

As noted in previous examples, it is wise to retain more than the number of significant figures while calculating the least-squares line, and to round off at the end of the computations. When a calculating machine is not being used, it is convenient to reduce the magnitudes of the x and y values by subtracting from the original values constants close in magnitude to $\bar{x}$ and $\bar{y}$ respectively. Subsequent readjustment may then be made near the end of the computations.

Problems

2–1   Estimate p and dp/dt for $t = 0$ from the data listed in Table 2–6. Compare the values so estimated from the original data with those estimated from the smoothed data.

2–2   From the values of C listed in Table 2–3 for $t = 20, 40, 50, 80, 90$, estimate C and $-dC/dt$ for $t = 60$.

2–3   Smooth the data for y listed in Table 2–2.

2–4   What polynomial equation of the form,

$$C = a + bt + ct^2 + \cdots$$

would be appropriate for the data listed in Table 2–3? Using the method of selected points fit the data to the chosen equation.

2–5   For the chemical reaction corresponding to the data listed in Table 2–6, the partial pressure of the reacting substance equals $2p_0 - p$, when p_0 represents the initial pressure (this may be taken as p for $t = 0$).

(a) Using the method of averages fit the original data to an equation of the form,

$$\log(2p_0 - p) = \log p_0 - kt/2\cdot303$$

and estimate the value of the first-order rate coefficient, k, in the units second^{-1}.

(b) Fit the data, by the simplified least-squares method, to the equation given in part (a). Estimate the values of k and p_0 together with their corresponding 90% confidence limits.

2–6   Using Eqn (2–29), construct a table similar to Table 2–5, illustrating the propagation of error in the estimation of the rate coefficient for a substance decomposing in solution by the second-order kinetic law,

$$k = (1/t)(1/C - 1/C_0)$$

As in Table 2–6, assume that the reaction has a half-life of 1000 s, that $\delta t = \pm 1$ s, that C is determined by titration with a maximum error of $\pm 0\cdot1$ ml, and that the original titre equals $20\cdot0$ ml for C_0.

2–7   Using Eqn (2–27), estimate the standard error in k at the half-life for the reaction system outlined in Problem 2–6. Assume that the standard errors in t, C_0, and C may be taken as two-thirds of the estimated maximum errors of reading as given in Problem 2–6.

2–8   Table 2–10 lists the rates of flow for a gas passing through a tubular reactor at 100 °C and 1·00 atm pressure. Estimate the total moles of gas which have passed through the reactor in one hour.

Table 2–10   Flow of gas through a tubular reactor

(See Problem 2–8)
V = volume of gas in millilitres at 100 °C and 1·00 atm;
t = time of flow in minutes.

t	dV/dt	t	dV/dt
0	60·0	35	66·1
5	75·9	40	50·0
10	83·2	45	36·7
15	86·0	50	27·9
20	86·2	55	21·5
25	84·1	60	17·6
30	78·8		

References

1. Lark, P. D., B. R. Craven, and R. C. L. Bosworth. *The Handling of Chemical Data.* Pergamon Press, Oxford, 1968.
2. Worthing, A. G., and J. Geffner. *The Treatment of Experimental Data.* John Wiley, New York, 1943.
3. Davis, D. S. *Nomography and Empirical Equations.* Reinhold, New York, 2nd ed., 1962.
4. Acton, F. S. *Analysis of Straight-Line Data.* John Wiley, New York, 1959.
5. Butler, R., and E. Kerr. *Introduction to Numerical Methods.* Pitman, London, 1962.
6. Booth, A. D. *Numerical Methods.* Butterworths, London, 2nd ed., 1957.
7. Guest, P. G. *Numerical Methods of Fitting.* Cambridge University Press, London, 1961.

3 Law and order of chemical change

The study of chemical kinetics involves an attempt to discover the laws of motion of chemical systems, or the changes that these systems undergo with time. Of prime importance is the change in the chemical composition of the system with time, but, in a wider sense, other types of motion may also be involved, such as the flow of matter into or out of the system, or the flow of heat through the system. Chemical engineering practice frequently demands an evaluation of all types of motion, but the laboratory study of chemical kinetics is usually aimed at the first, and experiments are commonly designed to maintain the other types of motion in a controlled or predictable condition.

Much of the basic work in chemical kinetics has been with 'closed' systems which are homogeneous and isothermal, that is, those in which matter does not flow in or out, and in which concentration or temperature gradients may be assumed insignificant, the temperature remaining substantially constant during the course of the chemical change. These are the systems which are the main concern of this chapter; 'open' systems, which involve the continual loss and/or gain of material, are discussed in Chapter 5.

3–1 Rate and stoichiometry

In the study of any chemical kinetic system, accurate information should first be sought about the material balance between those substances which disappear and those substances which emerge during the process of chemical change. Evidence about even trace substances is important as it could provide a key to a possible mechanism for the change.

Graphs of material balance plotted against reaction time, such as those displayed in Figs. 3–1 to 3–5, are particularly useful for identifying general features of a reaction. One of the simplest forms of behaviour is shown in Fig. 3–1. In this case, the reaction velocity, as evidenced by the rate of disappearance of reactant, is greatest at the commencement of the chemical change, and reduces continuously with time, the amount of reactant eventually becoming diminishingly small. In a complementary fashion, the rate of product formation is greatest at the start, and continually decreases as the reaction proceeds, a material balance between reactant and product being maintained at all times. The gas-phase conversion of cyclopropane to propene near 500 °C provides an example of this type of behaviour. As a variation, the amount of reactant remaining may approach a limiting equilibrium value which is significantly greater than zero (Fig. 3–2); that is, there is a significant opposing reaction. Many isomerization reactions of organic compounds behave in this way; examples are the interconversion of *keto* to *enol* forms, *cis-* to *trans-* forms, and *d-* to *l-* optical isomeric forms.

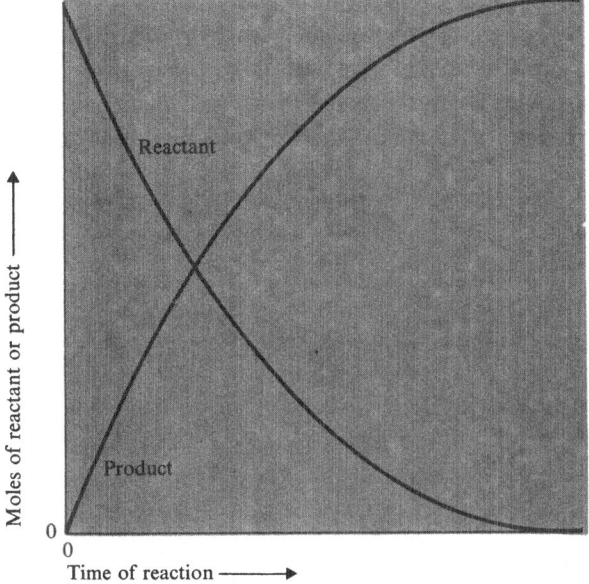

Fig. 3–1   Simple material balance plot for a reaction apparently going to completion.

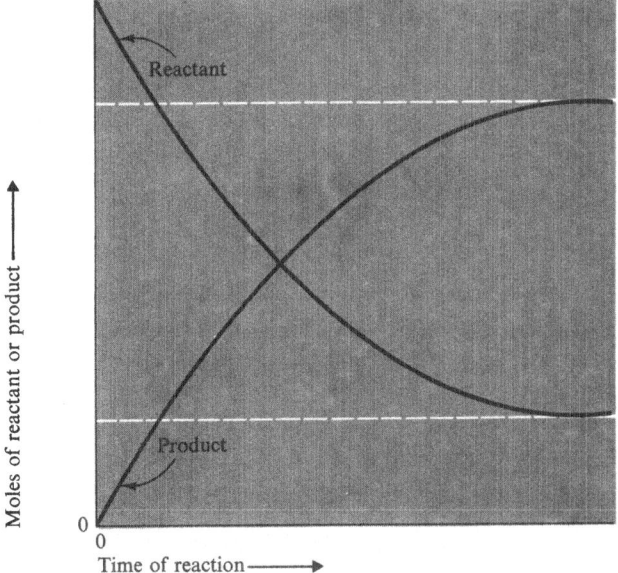

Fig. 3–2   Material balance plot for a reaction approaching a significant equilibrium.

Sometimes a false impression of approach to an equilibrium state is given by the behaviour of a reaction system which is self-inhibiting, or in which the reaction velocity has been prematurely reduced to a very low value by the progressive poisoning or destruction of a catalyst. It is advisable to check the validity of a suspected equilibrium state by attempting to approach it also from the 'product' side, that is, via the 'reverse' reaction. Further evidence may be provided by noting whether a predictable change in the equilibrium state is produced by the addition of one of the products of the reaction. Thermodynamic calculations are often useful in clarifying a situation, when suitable data are available.

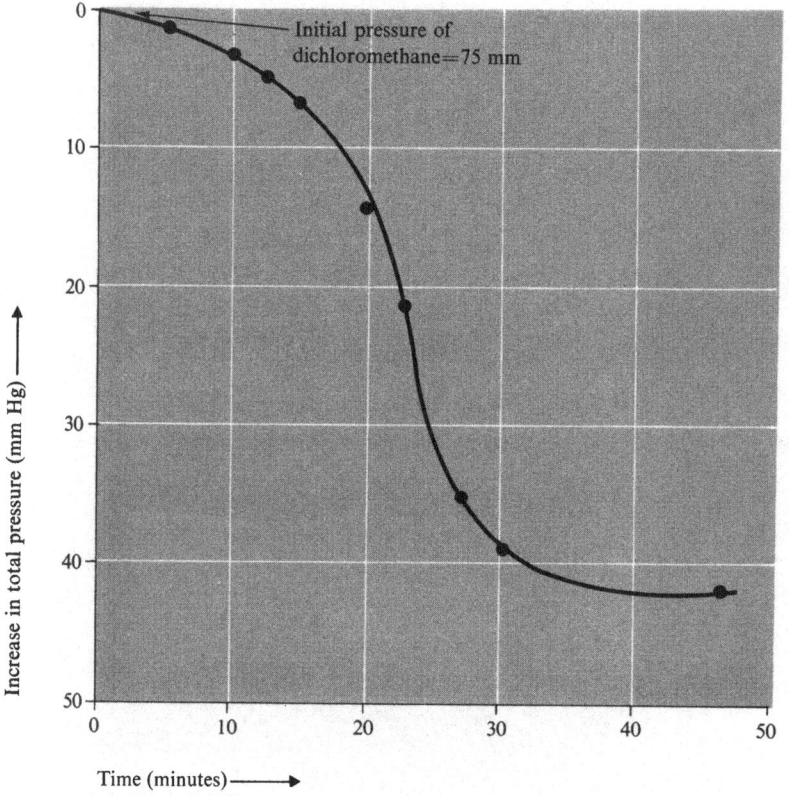

Fig. 3–3   Reaction behaviour characteristic of autocatalysis: the thermal decomposition of dichloromethane at 533 °C. The main products are hydrogen chloride and carbon; the increase in total pressure approximately corresponds to the pressure of dichloromethane which has decomposed. (M. R. Hoare, R. G. W. Norrish, and G. Whittingham, *Proc. Roy. Soc.* 1959, **A, 250**, 180.)

Figure 3–3 illustrates a case in which the rate of a chemical reaction, which is initially slow, increases to a maximum and eventually decreases once more. This type of behaviour is usually indicative of autocatalysis: the formation of a product which acts as a catalyst for the reaction. A

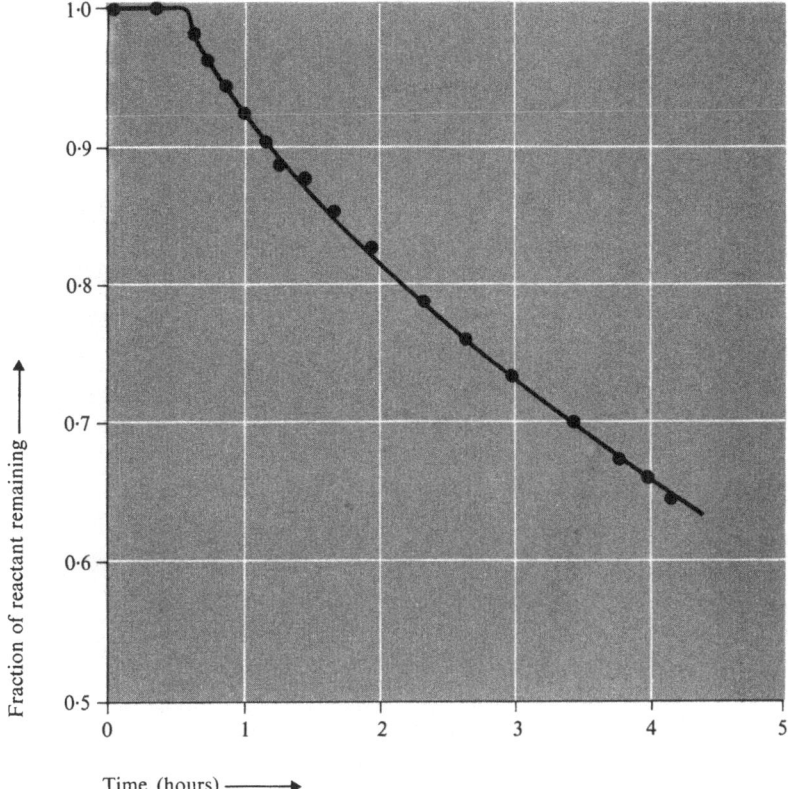

Fig. 3–4 Illustrating an induction period: the thermal decomposition of 1,2-dichloro-ethane at 400 °C (D. H. R. Barton and K. E. Howlett, *J. Chem. Soc.* 1949, 155).

simple example is the hydrolysis of an ester in near-neutral aqueous solution. The acid formed as a product acts as a catalyst for the hydrolysis which thus continues to increase in reaction velocity up to a saturation limit, beyond which the formation of further catalyst has little effect upon the rate, or the amount of ester becomes depleted, or the rate of the reverse (esterification) reaction becomes significant. In autocatalysis the time required for the system to attain maximum reaction rate is sometimes called the 'incubation' period.

The existence of an induction period is illustrated in Fig. 3–4. The reaction initially proceeds so slowly that its progress is virtually unobservable; during this time there is presumably a build-up towards a small but critical concentration of a reactive intermediate which suddenly triggers off the main reaction. Induction periods are common in radical-chain reactions.

A material balance graph of the kind illustrated in Fig. 3–5 is indicative of the occurrence of consecutive reactions: a product, such as *B* in the figure, undergoes subsequent chemical reaction to form a new product, *C*.

For example

$A \to B \to C$

A typical case is the hydrolysis of an organic nitrile in excess base, first to an amide which then converts to a carboxylate salt.

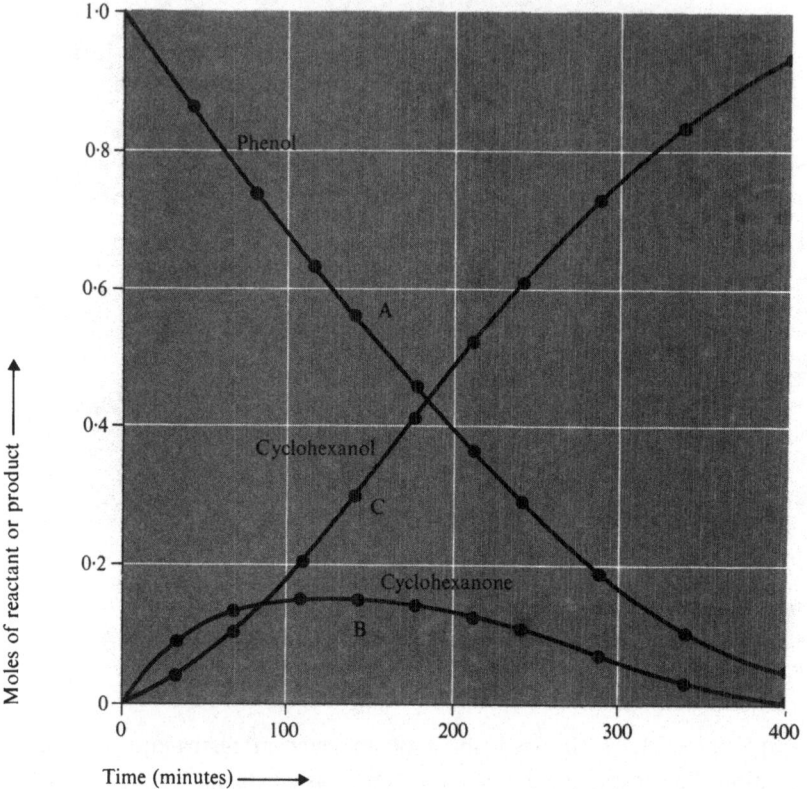

Time (minutes) ⟶

Fig. 3–5 Consecutive reactions: the hydrogenation of phenol on a nickel catalyst at 140 °C to cyclohexanol via cyclohexanone as intermediate (F. Coussemant and J. C. Jungers, *Bull. soc. chim. Belges*, 1950, **59**, 295).

The value of graphs showing the variation in material balance with time, therefore lies not only in their providing a quantitative picture of the fate of the chemical species present in the system, but also in their providing a guide for the classification of reaction behaviour in terms of rate and other factors. The graphs may alter substantially with changes in the reaction conditions, and assessment of these possible alterations provides further evidence for classification. The dependence of the rate of reaction upon the concentrations of the reacting species is of particular importance in this regard, but so also are the possible changes in the material balance graphs resulting from an alteration in the temperature or in the nature and extent of the surface of the reaction vessel, or from the addition of a suspected catalyst or inhibitor. See Example 3–1 at the end of the chapter.

The *rate of a chemical change* may be assessed in terms of the moles of reactant disappearing per unit time, $-dN_{reactant}/dt$, or the moles of product appearing per unit time, $+dN_{product}/dt$. For constant-volume systems, representation in terms of concentration is preferred to that in terms of moles, as the former is an intensive rather than an extensive measure. In either representation, however, the rate so defined is not a general one for the whole chemical change, but is defined for a particular chemical species in the system. Thus for the chemical change,

$$N_2 + 3H_2 \rightarrow 2NH_3$$

the relative rates for nitrogen, hydrogen, and ammonia are determined by the stoichiometry of the system, namely mole for mole, ammonia is always formed twice as fast as nitrogen disappears, and hydrogen always disappears three times as rapidly as does nitrogen. The numerical magnitude of the rate of chemical change depends upon whether it is expressed in terms of nitrogen, hydrogen, or ammonia. Rate coefficients are similarly affected. Because of this, it has been proposed by some that, for constant-volume systems, an unambiguous definition of the rate of chemical change is given by the rate of change in concentration of the particular species divided by the stoichiometric coefficient for the species in the chemical equation representing the change. Thus for the above example,

$$\text{rate of chemical change} = -dC_{N_2}/dt$$
$$= (-dC_{H_2}/dt)/3$$
$$= (+dC_{NH_3}/dt)/2$$

There is, as yet, no generally accepted convention for representing the rate of a chemical change. The 'unambiguous definition' referred to above is clear enough for changes which may be expressed in terms of chemical equations containing simple numerical stoichiometric coefficients, but it becomes unwieldy when the chemical change involves a number of different species, and when the stoichiometric coefficients are non-integral. In the nitration of toluene, for example, the relative proportions of the various nitrotoluenes formed vary with changes in the reaction conditions; a different specific definition of the rate of chemical change would therefore be required for each set of conditions. Many chemical changes involve a complicated sequence of linked reaction steps, and while the above definition of rate might be appropriate for any individual step, there appears to be limited value in its application to the overall chemical change.

It is less confusing, at least in the exploratory stages of examining the kinetic data for a particular chemical change, to analyse the information directly in terms of the rates of change in concentration of particular species, rather than in terms of a defined rate for the overall chemical change.

3–2   Rate and order

One useful classification of reaction behaviour is in terms of the concept of *order*. The order, n, with respect to a particular participant species, A, may be defined by Eqn (3–1) in terms of C_A, the concentration of A, and its rate of change with respect to time:

$$-dC_A/dt = kC_A^n \tag{3–1}$$

When there are several participant species involved in the chemical change, the order should be strictly defined in terms of each of these participant species, for example

$$-dC_A/dt = kC_A^n C_B^m \tag{3–2}$$

The reaction, so defined in terms of its velocity, is of the nth order in A, of the mth order in B, and so on. The overall order is $n+m$.

Although not all chemical reaction systems can be conveniently classified in terms of the concept of order, the classification is particularly useful for constant-volume systems exhibiting a material balance behaviour of the type illustrated in Fig. 3–1. For this reason, consideration of this graph type is of particular relevance to this chapter. Other types of material balance behaviour, such as those illustrated in Figs. 3–2 to 3–5 may usually be regarded as combinations of simpler reaction types. These more complicated reaction systems are discussed in greater detail in Chapter 4.

The curvature of the graph type shown in Fig. 3–1 will, of course, vary with the order of the reaction. Figure 3–6 shows these changes in curvature for three substances decomposing respectively via a zero-order, a first-order, and a second-order law, each decomposition having the same half-life.

Representation of Eqn (3–1) in logarithmic form is given by Eqn (3–3).

$$\log(-dC_A/dt) = \log k + n \log C_A \tag{3–3}$$

For the decomposition of a single substance, A, therefore the order of the reaction may be obtained directly from the slope of the graph of $\log(-dC_A/dt)$ plotted against $\log C_A$. This method of determining order, known as the *van't Hoff method*, may be applied to concentration/time data such as that used to construct the material balance graphs shown in Figs. 3–1 and 3–6. Numerical values for the differential term may be obtained from one of the methods outlined in Chapter 2. Equation (2–18)

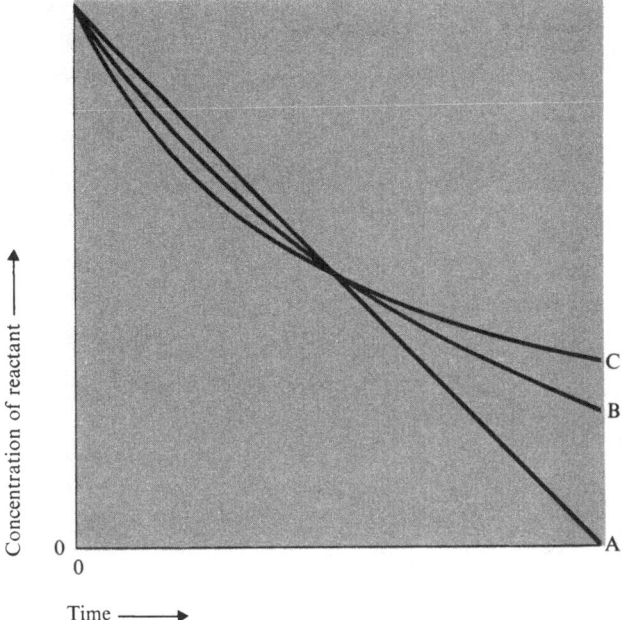

Fig. 3–6 Decompositions with a common half-life but proceeding by (A) zero-order,
(B) first-order and (C) second-order kinetic laws.

is of particular value in this regard, for it demonstrates that $dC/dt \approx \Delta C/\Delta t$
for the mid-point of the time interval, Δt.

Equation (3–3) may also be expressed in the approximate form of
Eqn (3–4), in which $\overline{C}$ represents the mid-point concentration over the
interval, ΔC.

$$\log(-\Delta C/\Delta t) \approx \log k + n \log \overline{C} \tag{3–4}$$

If the concentrations are measured at equal time intervals, Δt is constant,
and the simpler form, Eqn (3–5), is appropriate.

$$\log(-\Delta C) \approx \text{constant} + n \log \overline{C} \tag{3–5}$$

The order of a reaction may therefore be determined from the slope of
the graph of $\log(-\Delta C)$ plotted against $\log \overline{C}$. Simple variants of this
plot are possible. For example, in the gas-phase decomposition of cyclohexyl chloride to cyclohexene and hydrogen chloride, the concentration
of cyclohexyl chloride, expressed as a partial pressure p, is related to the
initial pressure p_0, and the total pressure p_T by the expression, $p = 2p_0 - p_T$.
The reaction order is therefore given directly by the slope of a graph of
$\log(\Delta p_T)$ plotted against $\log(2p_0 - \bar{p}_T)$. This is illustrated in Fig. 3–7
from data given in Table 3–1.

Reaction orders are rarely expressed in other than integral or simple
fractional values, because in these forms they may be more usefully

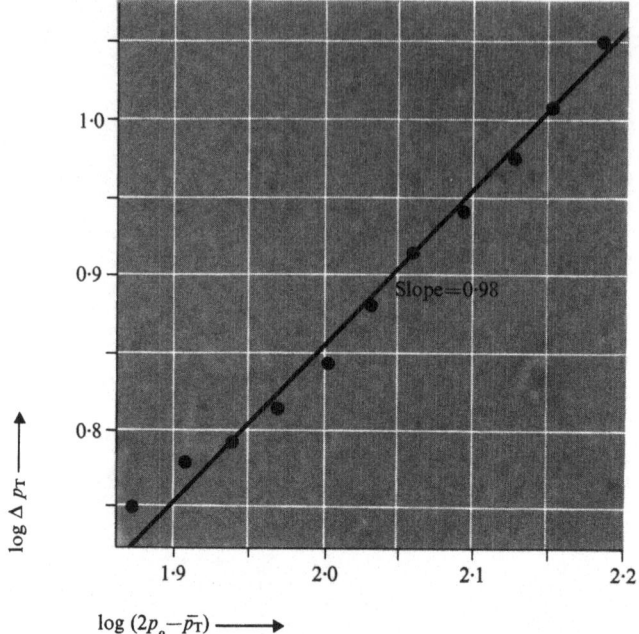

Fig. 3–7 Log-difference method for determining order with respect to time (see Table 3–1 for the data).

related to simple mechanistic models. Thus reactions for which the van't Hoff plots suggest orders of 0·9 or 1·1 would be classified as being of the first order. Orders of 0·5 and 1·5 are also acceptable, and are predictable in the case of certain classes of radical-chain reactions (see Chapter 5), however a reaction order such as 0·8 would suggest that the reaction is complex or that Eqn (3–1) does not represent a useful mathematical model for that particular chemical reaction. The choice of an alternative model is often a matter of trial and error, but an understanding of background physical theory can provide useful guidance. Thus, the decomposition of gaseous nitrous oxide on a heated platinum surface follows the rate law given by Eqn (3–6), the form of this law being suggested from Langmuir adsorption theory :*

$$-dp_{N_2O}/dt = ap_{N_2O}/(1+bp_{O_2}) \qquad (3–6)$$

An alternative procedure for determining the order of a reaction by the van't Hoff method consists of measuring the initial reaction rates for different initial concentrations of reactant, the logarithms of these rates being plotted against the logarithms of the corresponding initial concentrations. As before, the order is given by the slope of the graph (see

---

* See C. N. Hinshelwood, *The Kinetics of Chemical Change*, p. 192. Oxford, 1942.

Table 3–1 Log-difference method for determining order (van't Hoff)

(See Fig. 3–7)

Data are for the decomposition of gaseous cyclohexyl chloride at 334 °C, and represent the results of a single kinetic run. The reaction is followed by measuring the change in total pressure in the reaction vessel. t = time of reaction in minutes; p_T = total pressure (mm Hg); p_0 = initial pressure = 170·5 mm Hg.

t	p_T	Δp_T	$\log \Delta p_T$	$\overline{p_T}$	$\log(2p_0 - \overline{p_T})$
20	182·8				
		11·2	1·049	188·4	2·184
40	194·0				
		10·2	1·009	199·1	2·152
60	204·2				
		9·3	0·968	208·8	2·121
80	213·5				
		8·7	0·940	217·8	2·091
100	222·2				
		8·2	0·914	226·3	2·060
120	230·4				
		7·6	0·881	234·2	2·029
140	238·0				
		6·9	0·839	241·4	1·998
160	244·9				
		6·5	0·813	248·1	1·968
180	251·4				
		6·2	0·792	254·5	1·937
200	257·6				
		6·0	0·778	260·6	1·905
220	263·6				
		5·6	0·748	266·4	1·873
240	269·2				

Fig. 3–8 and Table 3–2). The French kineticist, Letort, has named this *the order with respect to concentration*, and that determined from the concentration versus time data *the order with respect to time*. For a given reaction, the two values for the order are not always identical. Of the two procedures, that based on initial rates is probably more realistic, since, in the initial stages of a reaction, there is less likelihood of interference by products which may act as catalysts or inhibitors for the chemical change.

When there is autocatalysis, the order with respect to time is less than the order with respect to concentration, because, during the course of a particular kinetic run, the drop in rate normally associated with the decreasing concentration of reactant is partially offset by the continued formation of catalyst which acts to raise the rate. Thus, in the gas-phase decomposition of neopentyl chloride, one of the products, hydrogen chloride, acts as a catalyst for the decomposition: the order with respect to time is near zero, and the order with respect to concentration is slightly greater than unity. Conversely, when there is auto-inhibition, the reaction

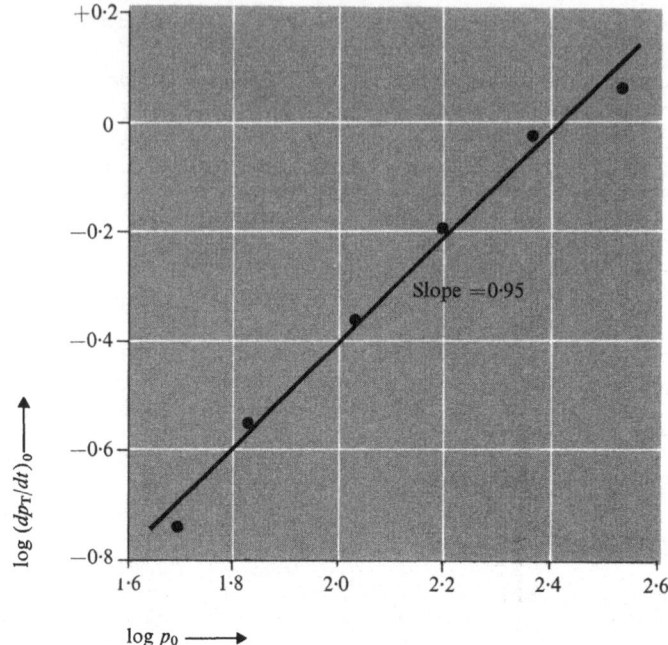

Fig. 3–8 Initial rate method for determining order with respect to concentration (see Table 3–2 for the data).

Table 3–2   Initial-rate method for determining order (van't Hoff)

(See Fig. 3–8)

Data are for the decomposition of gaseous cyclohexyl chloride at 334 °C, and represent the results of a number of separate kinetic runs with differing initial pressures of reactant (see also Table 3–1). t = time of reaction in minutes; p_T = total pressure (mm Hg); p_0 = initial pressure of reactant (mm Hg); $(dp_T/dt)_0$ = rate of change in total pressure at zero time (in mm min^{-1}) as estimated by the differentiated Gregory–Newton expression, Eqn (2–17).

	(Run 1)	(Run 2)	(Run 3)	(Run 4)	(Run 5)	(Run 6)
t	p_T	p_T	p_T	p_T	p_T	p_T
0	48·4	66·8	112·4	159·1	234·3	346·7
10	50·2	69·5	116·8	165·4	243·3	357·9
20	51·9	72·1	121·1	171·2	251·9	368·5
30	53·6	74·6	125·1	176·7	260·1	378·7
40	55·2	76·7	128·9	182·0	267·9	388·6
$(dp_T/dt)_0 =$	0·18	0·28	0·45	0·64	0·92	1·15
$\log(dp_T/dt)_0 =$	−0·745	−0·553	−0·347	−0·194	−0·036	+0·061
$\log p_0 =$	1·685	1·825	2·051	2·202	2·370	2·540

rate, during a particular kinetic run, decreases more rapidly than might be anticipated in terms of the decrease in reactant concentration alone, and the order with respect to time is greater than the order with respect to concentration. In the case of the thermal decomposition of acetaldehyde, for example, the two orders are respectively 2 and 1·5: some compound formed in the decomposition, possibly ethane, is tending to inhibit the reaction which is free radical in nature. The suspected catalytic or inhibitory action of products should be checked by carrying out kinetic runs with these substances added initially to the reactant. Examples and problems illustrating the differences between these two concepts of *reaction order* are given at the end of this chapter.

The *initial rate method* for determining order is also useful when there is more than one reactant. The rate of a reaction involving three reactants, A, B, and D, for example, may be expressed in terms of the following equations:

$$-dC_A/dt = kC_A^n C_B^m C_D^q \tag{3–7}$$

$$\log(-dC_A/dt) = \log k + n \log C_A + m \log C_B + q \log C_D \tag{3–8}$$

If, for a given series of kinetic runs, the initial concentrations of B and D are held constant and that of A is varied, the initial rate of the reaction $(-dC_A/dt)_0$ is related to $(C_A)_0$, the initial concentration of A by Eqn (3–9).

$$\log(-dC_A/dt)_0 = \text{constant} + n \log(C_A)_0 \tag{3–9}$$

The order, n, with respect to A is obtained from the slope of the graph of the logarithms of the initial rates plotted against the logarithms of the initial concentrations. In similar fashion the orders with respect to B and D may also be determined by two further series of kinetic runs.

Related to the initial rate method is the *isolation method* of Harcourt and Esson. In this method the concentrations of all species, except the one of interest, are held in excess. The concentrations of all reactant species (with the exception of this one) thus remain effectively constant during the course of the reaction, and the reaction behaviour with respect to the particular species of interest is therefore isolated from the remainder. The success of the method depends upon the one mechanism having the same dominant character over a wide range of concentrations of the reactants.

A commonly used method for determining order for a single reactant species is based upon the integrated form of the rate equation. This method therefore yields the order with respect to time. For a first-order reaction, $kt = \ln C_0 - \ln C$, and a graph of $\log C$ plotted against t should yield a straight line. In general, for a reaction of order n (provided $n \neq 1$),

$$kt = [1/(n-1)][1/C^{n-1} - 1/C_0^{n-1}] \tag{3–10}$$

and a graph of t against $1/C^{n-1}$ should yield a straight line. The method is one of trial and error, values of n being chosen until the data gives a satisfactory straight-line plot. Figure 3–9 shows the data from Table 3–1 plotted in this way, for $n = 0.5$, 1, and 1.5, and it is clear that, for this reaction, a first-order law is more closely obeyed than either of the other two. The method is not entirely satisfactory, as it determines the order by an indirect rather than by a direct procedure, and it depends upon a somewhat subjective assessment of 'goodness of fit' by the observer.

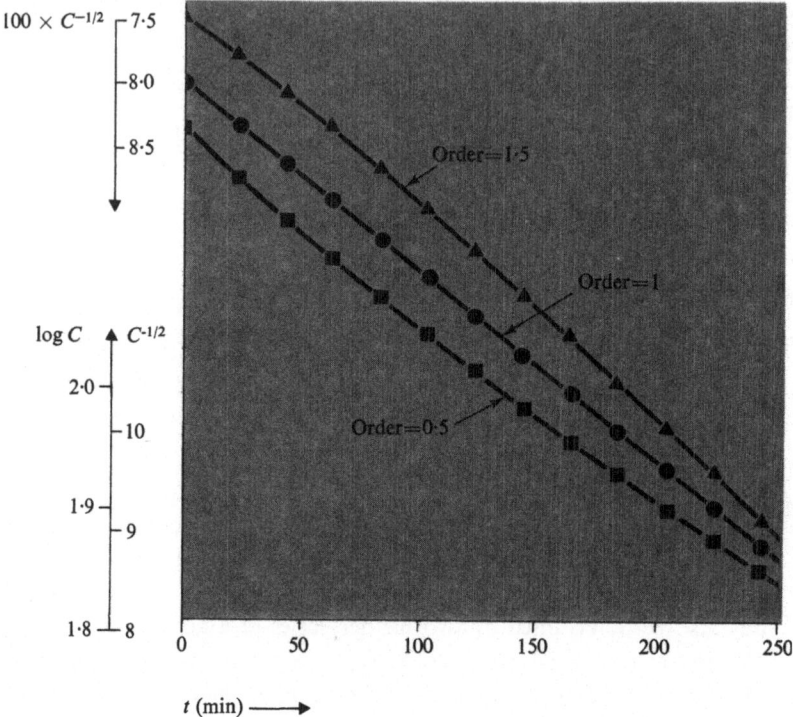

Fig. 3–9   Testing for order using the integrated form of the rate equation. Data are from Table 3–1. Concentration is expressed in mm. mercury pressure of gaseous reactant. Three different scales are used for the vertical axis.

A variation on this method, using dimensionless parameters, has been proposed by R. E. Powell[1] (pp. 14, 15). Equation (3–10) may be rewritten in the form of Eqn (3–11):

$$(C/C_0)^{1-n} - 1 = (n-1)kC_0^{n-1}t \tag{3 11}$$

The terms C/C_0 and $kC_0^{n-1}t$ are dimensionless, the latter being designated τ, a time parameter. Equation 3–11 therefore becomes

$$(C/C_0)^{1-n} - 1 = (n-1)\tau \tag{3-12}$$

and for the special case of the first-order equation,

$$\ln(C/C_0) = \tau \tag{3-13}$$

For a reaction of given order, n, there is a unique relationship between

(C/C_0) and τ, and the graphical forms of these relationships are shown in Fig. 3–10. In practice, experimental values of (C/C_0) are plotted against log t, since log τ cannot be determined until k and n are known. The graphs so obtained, however, have the same general shape as those for log τ, but they will be displaced along the log t axis. The value of n, the order of the reaction, is determined by comparison of the experimental curve with the theoretical shapes, such as those shown in Fig. 3–10. The *Powell plot* may be extended to reactions of complex behaviour, where it constitutes a most valuable identification method.

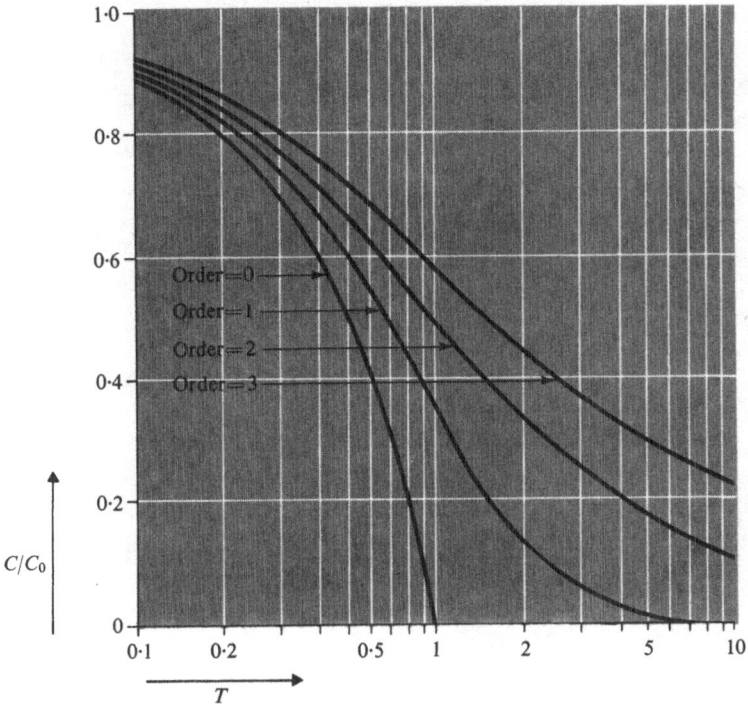

Fig. 3–10   The Powell plot for determining order (description is given in A. A. Frost and R. G. Pearson, pp. 14, 15).[1]

The *fractional-life method* is yet another procedure for determining order for a single reactant species. For a given fraction of reactant remaining, C/C_0, and given order, n, Eqn (3–11) may be expressed in the form of Eqn (3–14), in which t_f represents the time to reach this point of decomposition.

$$\log t_f = \text{constant} + (1 - n) \log C_0 \tag{3–14}$$

Equation (3–14) also holds for the special case of a reaction of the first order, for which,

$$\log t_f = \text{constant} \tag{3–15}$$

The value of $1-n$ is therefore determined from the slope of the graph of $\log t_f$ against the logarithm of the initial concentration ($\log C_0$). The method is applicable to a number of different runs with different initial concentractions or to time intervals within a single run, but it is subject to error if there is interference by the products in the course of a reaction.

When endeavouring to determine the order of a chemical reaction, allowance must be made for possible environmental effects upon the activities of the reactant species. These effects are of minor consequence for reactions in the gas phase (except at high pressure), or for molecular reactants in dilute solutions involving a single solvent. They are of somewhat greater consequence for molecular reactions in concentrated solutions or in solutions involving a range of solvents, and of major consequence for ionic reactions.

In the case of ionic reactions, for example, the influence of all ion-species in the solution is of importance, whether or not they are directly involved in the chemical reaction. This influence may be accounted for in terms of the ionic strength, u, of the solution, defined by Eqn (3–16), in which z_i represents the numerical charge (valency) of an ion, and C_i its concentration, the summation being made for all ion-species in the solution.

$$u = (1/2) \sum_i z_i^2 C_i \tag{3–16}$$

For a chemical change in dilute aqueous solution at 25 °C for which the critical step involves the reaction of two ion-species, A and B, theory predicts that for ionic strength u, the rate coefficient, k, is related to the rate coefficient, k_0, for infinitely dilute conditions, according to Eqn (3–17).

$$\log k = \log k_0 + 1 \cdot 02 \, z_A z_B \sqrt{u} \tag{3–17}$$

This equation (the Brönsted-Bjerrum equation), which predicts a straight-line plot of $\log k$ (or $\log [k/k_0]$) against the square root of ionic strength, has been tested successfully for a number of ionic reactions (see Fig. 3–11). The slope of the straight line should very nearly equal the product of the numerical charges on the participant ions. For further information about environmental effects upon reaction velocity, reference should be made to the textbooks listed in the References at the end of this chapter.

3–3   Rate, order, and the mass action law

It is often mistakenly supposed that the particular form of the rate equation may be predicted for a chemical reaction directly from its stoichiometry. It must be emphasized, however, that the appropriate rate equation (and the order of a reaction) can only be determined by experi-

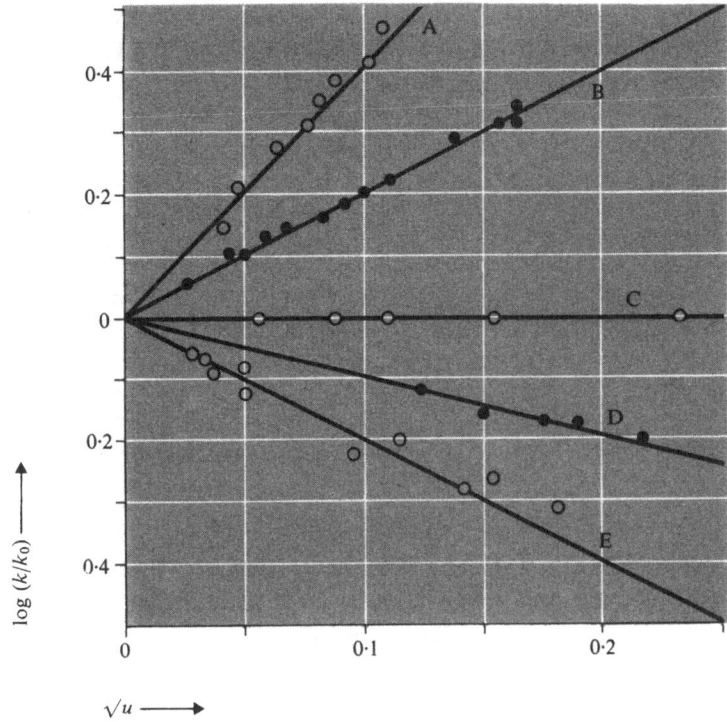

Fig. 3–11 Illustrating the effect of ionic strength, u, upon the rate coefficients of reactions involving ions. The graphs are drawn with slopes equal to integral values. Reactants: (A) $Co(NH_3)_5Br^{2+} + Hg^{2+}$; (B) $S_2O_8^{2-} + I^-$; (C) $CH_3COOC_2H_5 + OH^-$; (D) $H^+ + Br^- + H_2O_2$; (E) $Co(NH_3)_5Br^{2+} + OH^-$.

ment, and that there is no sure way of its prediction directly from an overall chemical equation.

The distinction between order and stoichiometry is well illustrated by the behaviour of the reaction between acetone and iodine in aqueous acid solutions; this reaction is represented by the following overall chemical equation:

$$CH_3COCH_3 + I_2 = CH_3COCH_2I + HI$$

On the basis of the mass action law, one may be tempted to predict the following form of the rate equation for this chemical reaction:

$$-dC_{I_2}/dt = kC_{\text{acetone}}C_{I_2} \tag{3–18}$$

Experiment has shown the reaction to be of the first order in acetone, but, over a wide range of concentrations, the reaction is of zero order in iodine and of the first order in hydrogen ion (see Eqn (3–19)), even though the latter species does not appear in the overall chemical equation.

$$-dC_{I_2}/dt = kC_{\text{acetone}}C_{H^+} \tag{3–19}$$

It is believed that the reaction proceeds by a sequence of steps, the first of these being the conversion of acetone from the *keto* to the *enol* form, this process being acid catalysed:

$$
\underset{(keto\ form)}{\overset{\displaystyle O}{\overset{\displaystyle \|}{CH_3.C.CH_3}}} + H^+ \rightleftharpoons \overset{\displaystyle OH^+}{\overset{\displaystyle |}{CH_3.C.CH_3}} \rightleftharpoons \underset{(enol\ form)}{\overset{\displaystyle OH}{\overset{\displaystyle |}{CH_3C}}} = CH_2 + H^+
$$

Addition of iodine to the *enol* form, and elimination of hydrogen iodide would then result in the formation of iodoacetone:

$$
\overset{\displaystyle OH}{\overset{\displaystyle |}{CH_3.C}} = CH_2 + I_2 \rightarrow \underset{\underset{\displaystyle I}{\displaystyle |}}{\overset{\displaystyle OH}{\overset{\displaystyle |}{CH_3C}}} - CH_2I \rightarrow \overset{\displaystyle O}{\overset{\displaystyle \|}{CH_3.C.CH_2I}} + HI
$$

If the *keto* → *enol* rearrangement is slow, relative to the rates of the steps involving iodine, then the overall rate of reaction depends upon the rate of production of the *enol* form, and this, in turn, depends upon the concentrations of acetone and hydrogen ion, and is independent of the iodine concentration. This is an example of a *rate-determining step* which governs the character of the overall reaction.

Modern acceptance of the mass action principle is in its application to an individual step of a chemical reaction rather than to the overall process. When the reaction consists of a number of steps, as in the last example, the experimentally deduced form of the rate equation may assist in the identification of a possible rate-determining step. A further illustration of this approach is given in Problem 3–3 at the end of the chapter.

It is important to realize that the particular form of the rate equation which applies in the early stages of a chemical reaction is not necessarily the same as that for equilibrium conditions. One must be cautious, therefore, in accepting a rate coefficient or a rate equation for a reverse reaction which has been deduced from the position of equilibrium and from the kinetic behaviour of a forward reaction. Deductions of this type, which often arise from erroneous application of the *principle of microscopic reversibility*, are only strictly valid under near-equilibrium conditions.

3–4   Rate and temperature

For most chemical reactions, the rate increases with temperature in the manner shown in Fig. 3–12, and the behaviour closely follows the form of the Arrhenius law given by Eqn (3–20):

$$\ln k = \ln A - E/RT \tag{3-20}$$

A plot of $\log k$ against $1/T$ should therefore approximate to a straight line, provided A and E are constant; this is illustrated by Fig. 3–13 for data on the gas-phase pyrolysis of 1,1-diethylcyclopropane (H. M. Frey and D. C. Marshall, *J. Chem. Soc.*, 1965, 191). Numerical methods for estimating A and E are given in Chapter 2 (see Examples 2–8 and 2–9).

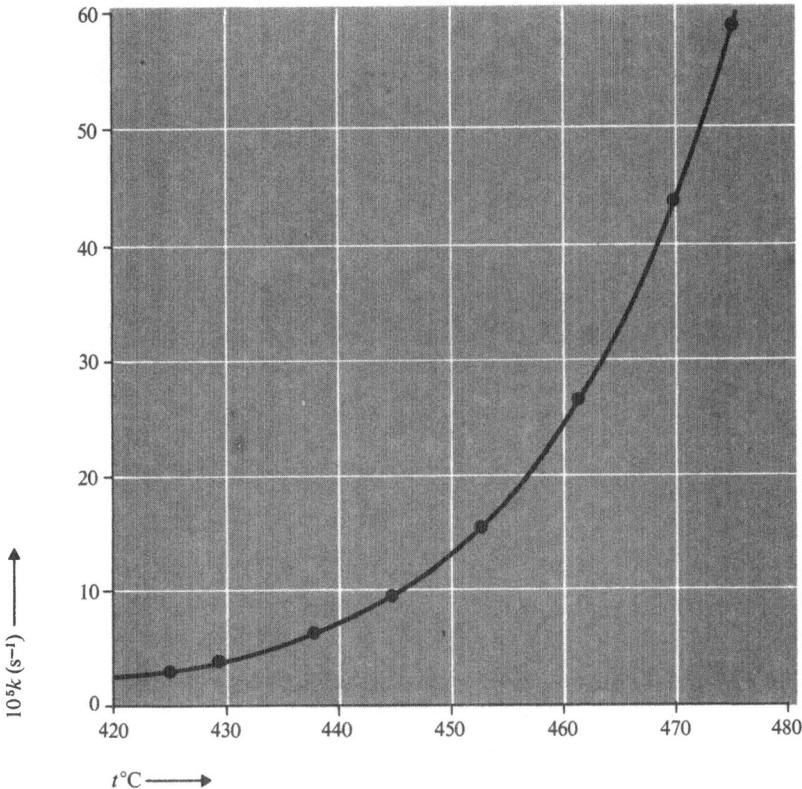

Fig. 3–12 The effect of temperature on reaction rate. Data are for the first-order rate coefficients for the gas-phase pyrolysis of 1,1-diethylcyclopropane.

Accurate observations on well-behaved rate data often disclose a slight but significant departure from a straight line when $\log k$ is plotted against $1/T$, but it is not possible, from such experimental data, to relate this departure unequivocally to lack of constancy in A or E. Reaction rate theories predict differing dependencies of A upon temperature: kinetic-molecular theory (bimolecular collisions) predicts that $A \propto T^{1/2}$, and the transition-state theory predicts that $A \propto T$. This suggests that a plot of either $\log k/T^{1/2}$ or $\log k/T$ against $1/T$ may lead to an improved linearity over the simple Arrhenius plot of $\log k$ against $1/T$. There is, however, no practical evidence of any one of these plots having

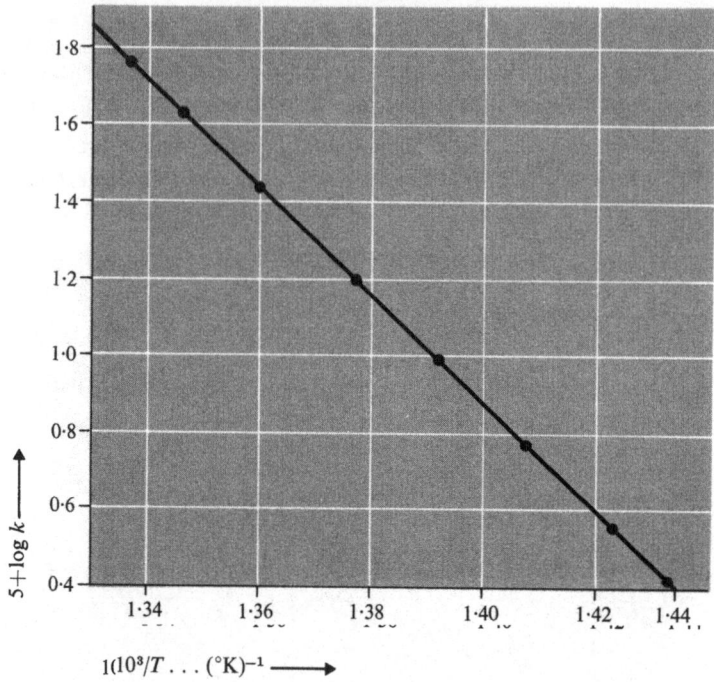

Fig. 3–13   Arrhenius plot for the temperature dependence of the rate of pyrolysis of 1,1-diethylcyclopropane (see also Fig. 3–12).

a particular advantage over the other two, as regards linearity, when applied to a wide range of experimental data, and, for this reason, the simple Arrhenius plot of $\log k$ against $1/T$ is usually preferred.

Major departure of data from the Arrhenius law usually indicates a complex mechanism or a change in the dominant mechanism over the range of recorded data. Figures 3–14, 3–15, and 3–16 provide examples of this. The graph shown in Fig. 3–14 could result from the occurrence of two concurrent reactions with widely different activation energies, one reaction (for example, a homogeneous process) being dominant at higher temperatures and the other reaction (a heterogeneous process) being dominant at lower temperatures (see Problem 3–7). In Fig. 3–15 the sudden discontinuity in the curve could represent an ignition temperature, above which a branched chain reaction occurs with explosive violence, and below which a milder molecular reaction takes place as the dominant process. A graph of the type shown in Fig. 3–16 may result from the decreasing efficiency of a catalyst or from its destruction with increase in temperature. In these cases the Arrhenius graph provides information which is of limited value quantitatively, but it may provide useful qualitative evidence about the complexity of the mechanism of chemical change.

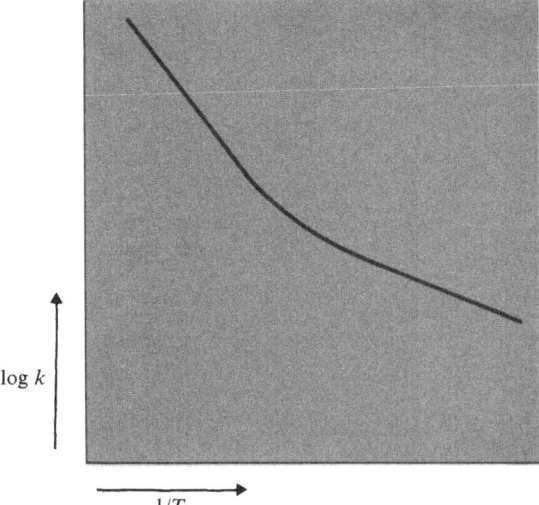

Fig. 3–14   Arrhenius plot for two concurrent reactions with widely different activation energies.

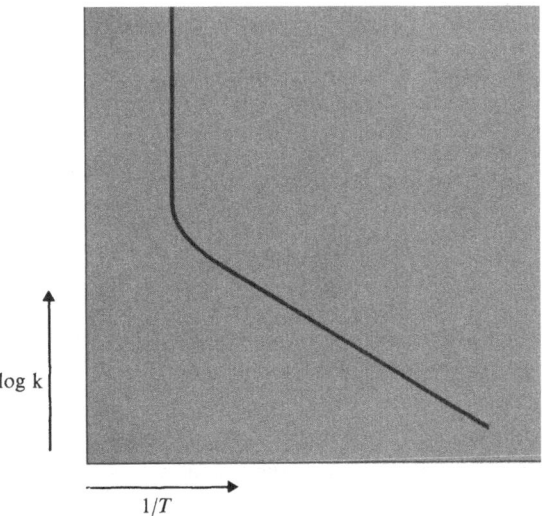

Fig. 3–15   Arrhenius plot showing an ignition temperature.

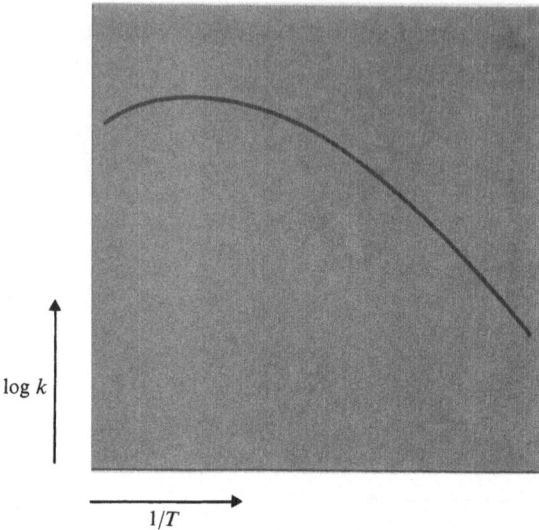

Fig. 3–16 Arrhenius plot typical of catalyst breakdown with increasing temperature.

Examples and problems

Example 3–1 Analysis of a complex chemical reaction

The following example illustrates the use of material balance graphs and of observations on reaction order for the clarification of the behaviour of a complex chemical reaction. The information is for the thermal decomposition of gaseous neopentyl chloride at 400–500 °C (J. S. Shapiro and E. S. Swinbourne, *Canadian J. Chem.*, 1968, **46**, 1341, 1351).

Figure 3–17 is a material balance graph for the products formed during the pyrolysis of neopentyl chloride at 444 °C. In the early stages of the pyrolysis, the main products are hydrogen chloride, and methylbutenes plus 1,1-dimethylcyclopropane (C_5H_{10} products); there are smaller amounts of methane, isobutene, methyl chloride, and 1-chloro-2-methylpropane also formed. The relative proportions of the C_5H_{10} products at different stages of the decomposition are illustrated in Fig. 3–18, together with their equilibrium proportions: it is clear that the C_5H_{10} products are formed in non-equilibrium proportions and that the 1,1-dimethylcyclopropane undergoes subsequent decomposition and/or isomerization as the main reaction proceeds. The shape of the material balance graph for 3-methylbut-1-ene suggests that this compound is not an initial product of the main reaction (Fig. 3–5), but could be formed from the subsequent isomerization of the 1,1-dimethylcyclopropane; separate studies on this latter reaction have supported this proposition.

Studies on the rate of decomposition of neopentyl chloride showed that the order with respect to time is close to zero (see Fig. 3–17) and that the order with respect to concentration is about 1·3, the difference between these two values indicating the occurrence of autocatalysis. When kinetic runs were carried out with substances added initially to the reactant, hydrogen chloride was found to be a catalyst, and olefinic compounds were found to have inhibitory effects upon the formation of all products except 2-methylbut-1-ene. Inhibitory effects of this nature are characteristic of radical-chain reactions, but the independent nature of the formation of 2-methylbut-1-ene suggests that it may be formed via a molecular mechanism. The catalytic action of hydrogen chloride could result from its reaction with radicals to produce chlorine atoms which are effective as chain-carriers.

The experimental evidence, therefore, appears to be consistent with a proposition that neopentyl chloride decomposes via two concurrent processes, one of which is molecular and the other radical chain. The detailed testing of this proposition involves careful and painstaking analysis of the experimental results, but the primary evidence for the proposition is provided by the nature of the material balance graphs.

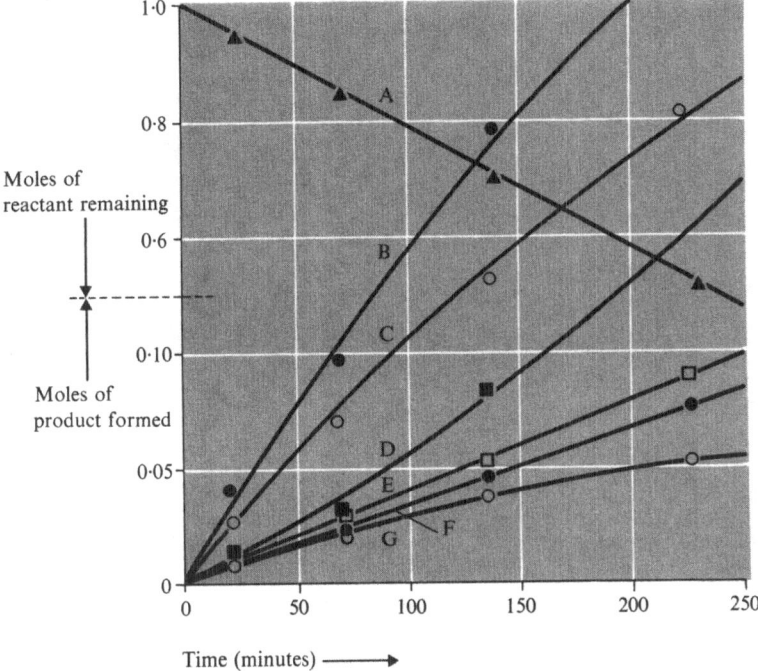

Fig. 3–17 Material balance graph for the decomposition of neopentyl chloride at 444 °C. (A) Neopentyl chloride; (B) Hydrogen chloride; (C) Methylbutenes plus 1,1-dimethcyclopropane; (D) Methane; (E) Isobutene; (F) Methyl chloride; (G) 1-chloro-2-methylpropene.

Problems

3–1 The data given in Table 3–3 are for the decomposition of aqueous hydrogen peroxide catalysed by ferric chloride. By means of appropriate graphs based on the integrated forms of the rate equations, determine whether the data more closely obey a first- or a second-order law.

3–2 Table 3–4 lists data for the hypothetical chemical reaction given by the following stoichiometric equation:

$$3A + 2B + D = 2E + F$$

(a) Express the rate of formation of F in terms of the rates of disappearance of A, B, and D.

(b) Express the rate of formation of F in terms of the concentrations of A, B, and D.

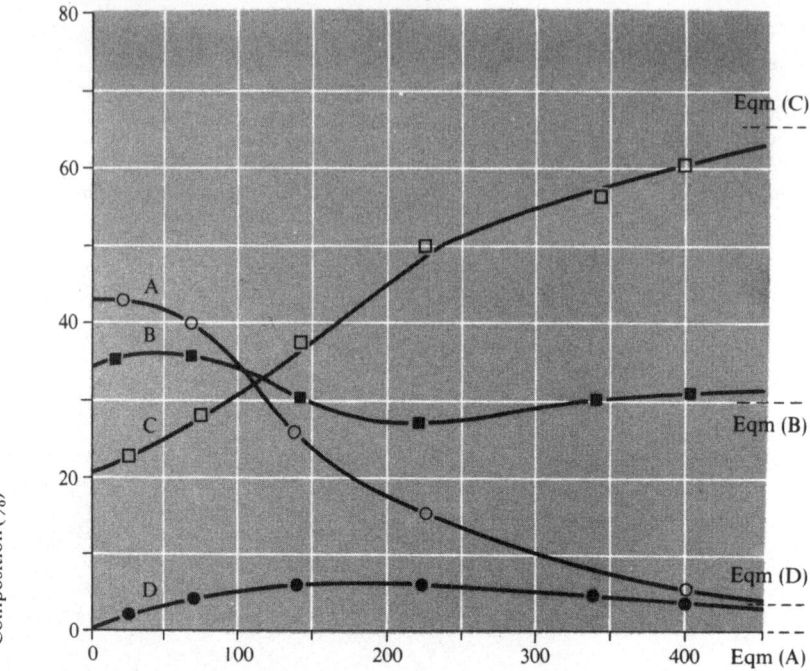

Fig. 3–18 Composition of C_5H_{10} products at different stages of the decomposition of neopentyl chloride. (A) 1,1-dimethylcyclopropane; (B) 2-methylbut-1-ene; (C) 2-methylbut-2-ene; (D) 3-methylbut-1-ene. The equalibrium composition is shown on the right.

Table 3–3   Decomposition of aqueous hydrogen peroxide

(See Problem 3–1)

The decomposition is catalysed by ferric chloride in aqueous solution at 25 °C. t = time of reaction in minutes; x = titration figure in millilitres when a 10 ml aliquot of the H_2O_2 solution is titrated with 0·02 molar $KMnO_4$ in dilute H_2SO_4.

t	x
0	61·9
10	55·6
20	50·4
30	45·9
40	42·0
50	38·4
60	35·1
70	31·9

3–3   The reaction of iodide ion with persulphate ion in aqueous acid solution is represented by the following stoichiometric equation:

$$2I^- + S_2O_8^{2-} = 2SO_4^{2-} + I_2$$

The initial rate of the chemical reaction may be evaluated from the time taken to form a known amount of iodine. In one series of experiments, the iodide-ion

Table 3–4   Rate data for a hypothetical chemical reaction

(See Problem 3–2)

C_{A_0}; C_{B_0}; C_{D_0} = initial concentrations of A, B, D in millimoles litre^{-1}; r_{F_0} = initial rate of formation of F in millimoles litre^{-1} min^{-1}.

Experiment no.	C_{A_0}	C_{B_0}	C_{D_0}	r_{F_0}
1	10	10	10	0·5
2	20	10	10	0·5
3	30	10	10	0·5
4	10	20	10	1·0
5	10	30	10	1·5
6	10	10	20	2·0
7	10	10	30	4·5

concentration was kept constant while the persulphate-ion concentration was varied, and, in a second series, the persulphate-ion concentration was kept constant while the iodide-ion concentration was kept constant; in both series the hydrogen-ion concentration was kept constant and so also was the ionic strength of the solution. The results of these two series of experiments are given in Table 3–5.

(a) What is the order of the reaction with respect to iodide ion and with respect to persulphate ion?

(b) It has been proposed that the reaction proceeds by the following consecutive steps:

$$I^- + S_2O_8^{2-} + H_2O \rightarrow 2SO_4^{2-} + HIO + H^+ \qquad \text{Step (1)}$$

$$H^+ + I^- + HIO \rightarrow H_2O + I_2 \qquad \text{Step (2)}$$

From your deductions in (a) which of these two steps would you judge to be rate-determining, if they correctly represent the reaction mechanism?

Table 3–5   Reaction of persulphate ion with iodide ion

(See Problem 3–3)

$(S_2O_8^{2-})_0$ = initial concentration of $S_2O_8^{2-}$ in (moles litre^{-1}) $\times 10^2$; $(I^-)_0$ = initial concentration of I^- in (moles litre^{-1}) $\times 10^2$; t = time in seconds for a known concentration of iodine to be formed.

Series 1 (fixed I^- concentration):							
$(S_2O_8^{2-})_0$	4·0	2·9	2·0	1·45	1·1	0·73	0·37
t	30	40	61	85	111	165	345

Series 2 (fixed $S_2O_8^{2-}$ concentration):							
I^-	8·0	5·8	4·0	2·9	2·1	1·45	0·73
t	31	46	58	72	92	120	180

3–4   Table 3–6 lists data for the rate coefficient for the reaction between aqueous persulphate ion and iodide ion at 25 °C in solutions of differing ionic strengths (C. V. King and M. B. Jacobs, *J. Amer. Chem. Soc.*, 1931, **53**, 1704).

(a) By means of a suitable graph based upon the Brönsted-Bjerrum equation, Eqn (3–17), estimate the limiting value of the rate coefficient at infinite dilution.

Table 3–6   Variation of rate coefficient with ionic strength—reaction of persulphate ion with iodide ion

(See Problem 3–4)

u = ionic strength of the solution (Eqn (3–16)); k = rate coefficient in litre mole^{-1} min^{-1}.

u	k
0·00062	0·095
0·00095	0·102
0·00125	0·105
0·00175	0·108
0·00250	0·111
0·00326	0·113
0·00350	0·114
0·00428	0·116
0·00503	0·123

(b) From the graph in (a) estimate the product of the numerical charges upon the pair of ions participating in the slow step of the reaction.

3–5   Gaseous acetaldehyde decomposes thermally to methane and carbon monoxide as the only major products.

$$CH_3CHO = CH_4 + CO$$

Table 3–7 shows the results of a kinetic study of the decomposition at 518 °C, the course of the decomposition having been followed by measuring the change in total pressure with time in a closed reaction system.

Determine the order of the reaction with respect to time
(a) by means of a log-difference plot (van't Hoff method);
(b) by means of the fractional-life method;
(c) by means of a Powell plot.

Table 3–7   Pyrolysis of acetaldehyde—data for a single kinetic run at 518 °C

(See Problem 3–5)

t = time of reaction in seconds; p = total pressure in reaction vessel (mm Hg).

t	p	t	p
0	360	800	601
100	430	900	610
200	477	1000	618
300	514	1100	625
400	542	1200	632
500	563	1300	638
600	578	1400	643
700	590	1500	647

3–6   Table 3–8 lists data for the initial rate of decomposition of gaseous acetaldehyde at 500 °C for a number of kinetic runs with differing initial pressures of reactant. Determine the order of the reaction with respect to concentration.

The first-order rate coefficients (k_{total}) for ethyl-chloride decomposition in a packed vessel at different temperatures are given in Table 3–9 together with the corresponding values of the rate coefficients for the homogeneous decomposition ($k_{homogeneous}$).

(a) Plot the values of log k_{total} against the reciprocal of the absolute temperature and compare the graph with Fig. 3–14.

(b) Estimate separately the values of the Arrhenius activation energy for the homogeneous decomposition and for the heterogeneous decomposition.

Table 3–8 Pyrolysis of acetaldehyde—initial rates for different initial pressures

(See Problem 3–6)

p_0 = initial pressure of acetaldehyde (mm Hg); $(dp/dt)_0$ = initial rate of total pressure in mm Hg s^{-1}.

p_0	$(dp/dt)_0$
84·7	0·048
116·4	0·076
158·1	0·119
212·8	0·180
257·0	0·272
363·1	0·377
433·5	0·557
532·1	0·716

3–7 Near 400 °C, ethyl chloride decomposes in a packed glass vessel to ethylene and hydrogen chloride by a combination of heterogeneous and homogeneous processes. The homogeneous decomposition may be studied separately in an unpacked glass vessel, the surface of which has been coated with a non-polar material.

Table 3–9 Pyrolysis of ethyl chloride

(See Problem 3–7)

t °C = temperature in degrees Centigrade; k_{total} = first-order rate coefficient (in s^{-1}) for the combined heterogeneous and homogeneous processes; $k_{homogeneous}$ = first-order rate coefficient (in s^{-1}) for the homogeneous decomposition.

t °C	$10^4 \times k_{total}$	$10^4 \times k_{homogeneous}$
380·3	0·268	0·023
390·5	0·417	0·047
403·8	0·581	0·112
413·4	0·855	0·204
427·4	1·42	0·486
431·3	1·70	0·636
439·1	2·21	1·03
447·5	3·28	1·61
461·3	5·39	3·61

References

1. Frost, A. A. and R. G. Pearson. *Kinetics and Mechanism*. John Wiley, New York, 2nd ed., 1961. See particularly Chapters 2 and 3.
2. Laidler, K. J. *Chemical Kinetics*. McGraw-Hill, New York, 2nd ed., 1965. See particularly Chapter 1.
3. Friess, S. L., E. S. Lewis, and A. Weissberger (Eds.). *Investigation of Rates and Mechanisms of Reactions* (vol. 8 of *Technique of Organic Chemistry*). John Wiley (Interscience), New York, 2nd ed., 1961. See particularly, Chapter 5 by R. Livingston, and Chapter 6 by J. F. Bunnet in Part 1.
4. Ashmore, P. G. *Principles of Reaction Kinetics*. R.I.C. Monographs for Teachers, No. 9.
5. Ashmore, P. G. In *Education in Chemistry*, 1965, **2**, 160.
6. Hougen, O. A. and K. M. Watson. *Kinetics and Catalysis* (Part 3 of *Chemical Process Principles*). John Wiley, New York, 1947.
7. Krupka, R. M., H. Kaplan, and K. J. Laidler. The kinetic consequences of the principle of microscopic reversibility. *Trans. Faraday Soc.*, 1966, **62**, 2754.

4 Estimation of rate coefficients

The choice of an appropriate method for estimating a rate coefficient is dictated largely by the behaviour of the reaction system and the experimental procedures adopted for its study. Reliable results are usually the outcome of sensible planning of experiments, and, in this respect material balance graphs, such as those outlined in Chapter 3, can provide valuable guidance to methods which are likely to prove successful.

In the case of reactions of simple order and stoichiometry for which data analysis involves the estimation of a single rate coefficient, a range of reliable and tested methods is available. Complex reaction systems are usually considered as combinations of simple reaction types, and as such the analysis of these systems involves the estimation of two or more rate coefficients; extraction of the required information is correspondingly more difficult in these cases, and the range of satisfactory methods is somewhat restricted. In all cases increased reliability may be attached to the estimated value of a rate coefficient when its constancy has been checked over an extended range of experimental conditions, and by the use of more than one computational method.

Attention in this chapter is focused upon selected methods which have been found reliable for simple and for moderately complex reactions proceeding at constant temperature in closed systems. The discussion is not intended to be exhaustive, but rather as an attempt made to set selected methods into perspective in order to build up an understanding of their relative merits, and to point the way towards projections of these methods which could prove of value in special cases. Increasing use is now being made of computer-based techniques, particularly in the analysis of the more complex reaction systems. It has not been considered convenient to attempt a discussion of these techniques in a text of this size, and interested readers should consult the References at the end of this chapter.

4–1 Use of integrated rate equations

A rate coefficient is defined in terms of a differential equation, the usual form of which expresses the rate of change in concentration of a particular chemical species as a function of the concentrations of species present in the system. (In the case of a photochemical change the light intensity is also involved in the rate equation.) In most kinetic studies, however, the *rate* is not measured directly, but rather the concentration of one or a number of species is measured at particular instants of time during the course of the reaction. Two forms of analysis are therefore evident: either an attempt can be made to evaluate the differential coefficients from the time-concentration data, or to integrate the differential equation so that a direct functional relationship between the variables, time and concentration, is achieved. As discussed in the last chapter, the integrated form of the rate equation is of greatest use for a well-behaved reaction following a simple order.

For a single chemical species disappearing according to the rate law, $-dC/dt = kC^n$, integration of the rate equation yields the following relationship for cases other than order, $n = 1$:

$$kt = \left(\frac{1}{n-1}\right)\left(\frac{1}{C^{n-1}} - \frac{1}{C_0^{n-1}}\right) \tag{4-1}$$

The integration is subject to the boundary condition that $C = C_0$ when $t = 0$. Equation (4-1) may be rewritten in the form of Eqn (4-2) from which it is evident that a graph of $1/C^{n-1}$, as abscissa, plotted against t, as ordinate, should be a straight line with slope equal to $(n-1)k$.

$$\frac{1}{C^{n-1}} = \frac{1}{C_0^{n-1}} + (n-1)kt \tag{4-2}$$

The corresponding relationships for the first-order case, $n = 1$, are given by Eqns (4-3) and (4-4), and the straight-line graph with slope equal to minus $k/2\cdot303$ is provided by a plot of $\log C$, as abscissa, against t, as ordinate.

$$kt = \ln C_0 - \ln C \tag{4-3}$$

$$\log C = \log C_0 - (k/2\cdot303)t \tag{4-4}$$

Plots of this nature have been traditionally used for estimating the rate coefficients for many well-behaved reaction systems. An example of a first-order plot for the decomposition of gaseous cyclohexyl chloride is given in Fig. 3-9; Fig. 4-1 shows a graph for the dimerization of 1,3-butadiene—a reaction which proceeds by a second-order kinetic law (see also Table 4-1). In all cases, the units of the rate coefficient, k, are $(\text{concentration})^{1-n}(\text{time})^{-1}$. As described in Chapter 2, application of the method of averages or the method of least-squares to the numerical data yields a more satisfactory estimate of the rate coefficient than that obtained from the direct estimate of slope from the graph itself.

When two chemical species, A and B, are reacting mole for mole according to the following rate law,

$$-dC_A/dt = -dC_B/dt = kC_A C_B \tag{4-5}$$

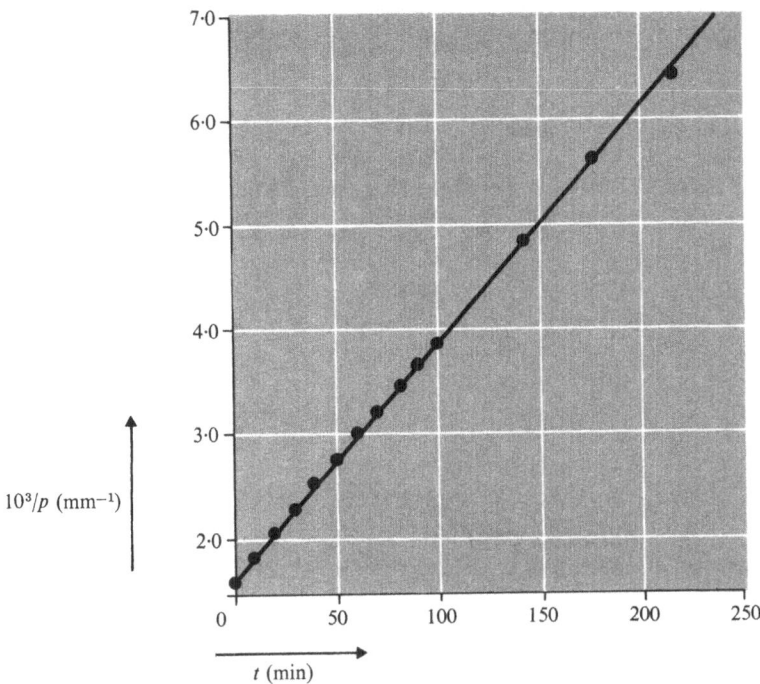

Fig. 4–1   Dimerization of 1,3-butadiene: a second-order plot using the integrated form of the rate equation. Data are from Table 4–1 (p represents the partial pressure of 1,3-butadiene in mm mercury).

integration yields Eqns (4–6) and (4–7) with boundary conditions, $C_A = C_{A_0}$ and $C_B = C_{B_0}$ for $t = 0$.

$$kt = \frac{1}{C_{A_0} - C_{B_0}} \ln \left| \frac{C_{B_0} C_A}{C_{A_0} C_B} \right| \tag{4–6}$$

$$\log \frac{C_A}{C_B} = -\log \frac{C_{B_0}}{C_{A_0}} + \left[k \left(\frac{C_{A_0} - C_{B_0}}{2 \cdot 303} \right) \right] t \tag{4–7}$$

A plot of $\log (C_A/C_B)$, as abscissa, against t, as ordinate, gives a straight line with slope equal to $k(C_{A_0} - C_{B_0})/2 \cdot 303$.

A graph of this type for the mixed second-order reaction between sodium ethoxide and ethyl dimethyl sulphonium iodide in ethanol is shown in Fig. 4–2 (see also Table 4–2).

Equations (4–6) and (4–7) hold only for $C_{A_0} \neq C_{B_0}$; if these two initial concentrations are exactly equal, the situation is mathematically equivalent to a reaction of the second order involving one reactant, with the integrated form of the rate equation given by Eqn (4–8).

$$kt = 1/C_A - 1/C_{A_0} = 1/C_B - 1/C_{B_0} \tag{4–8}$$

Table 4–1 Dimerization of 1,3-butadiene

(See Fig. 4–1)

Data are for the dimerization of gaseous 1,3-butadiene to vinyl cyclohexene at 326 °C. (Interpolated from the results of W. E. Vaughan, *J. Amer. Chem. Soc.*, 1932, **54**, 3863.) t = time of reaction in minutes; p_T = total pressure (mm Hg) in reaction vessel; p_0 = initial pressure (mm Hg) of reactant; p = partial pressure (mm Hg) of 1,3-butadiene ($p = 2p_T - p_0$).

t	p_T	p	$(1/p) \times 10^3$
0	(634·1)*	(634·1)	(1·577)
5	611·0	590·0	1·695
10	591·9	551·8	1·812
15	573·7	515·4	1·940
20	559·1	486·2	2·057
25	545·7	459·4	2·177
30	533·6	435·2	2·297
35	523·9	415·8	2·405
40	513·7	395·4	2·529
50	497·4	362·8	2·756
60	483·9	335·8	2·978
70	427·5	313·0	3·195
80	462·4	292·8	3·415
90	453·4	274·8	3·639
100	445·2	258·4	3·870
120	432·2	232·4	4·302
140	420·5	209·0	4·785
180	404·1	176·2	5·675
220	393·3	154·6	6·468
260	380·9	127·7	7·830

*Estimated by extrapolation.

When the initial concentrations are nearly equal, it is convenient to use another form of equation involving C_M, the mean concentration of A and B at time, t, and C_S, one-half of the excess concentration of A over B. Thus,

$$C_A = C_M + C_S \qquad (4–9)$$

and

$$C_B = C_M - C_S \qquad (4–10)$$

Substitution in Eqn (4–5) gives,

$$-dC_A/dt = k(C_M + C_S)(C_M - C_S)$$

and

$$-dC_M/dt = k(C_M^2 - C_S^2)$$
$$= kC_M^2[1 - (C_S/C_M)^2] \qquad (4–11)$$

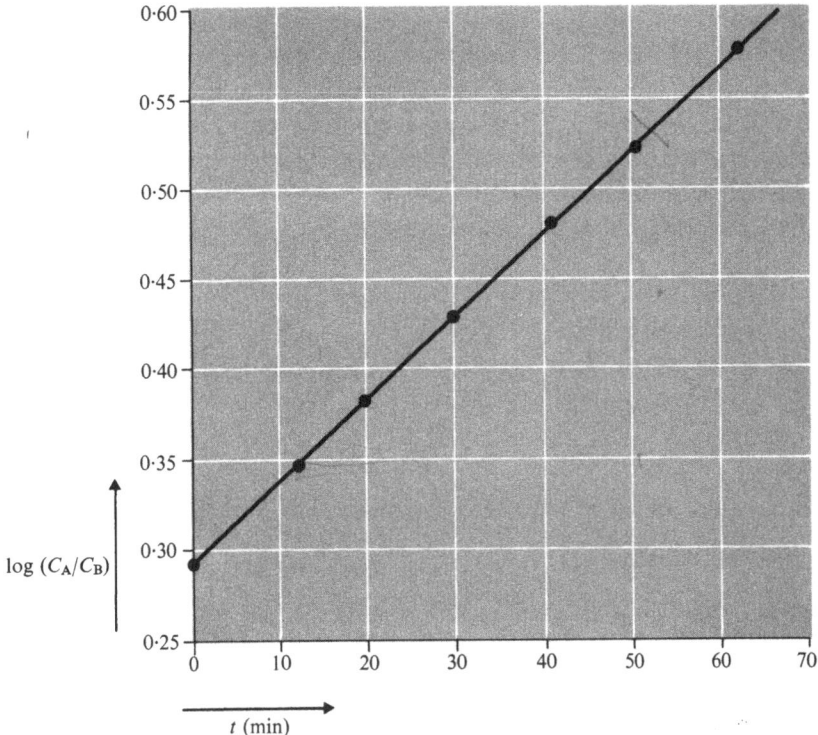

t (min)

Fig. 4–2 Reaction between sodium ethoxide and ethyl dimethyl sulphonium iodide: a second-order plot using the integrated form of the rate equation. Data are from Table 4–2.

Table 4–2   Example of a mixed second-order reaction

The data are for the reaction between sodium ethoxide and ethyl dimethyl sulphonium iodide in ethanol at $64.0\,°C$ (E. D. Hughes, C. K. Ingold, and G. A. Maw, *J. Chem. Soc.*, 1948, 2072).

$$NaOC_2H_5 + (C_2H_5)(CH_3)_2SI = \begin{cases} NaI + (C_2H_5)_2O + (CH_3)_2S \\ \text{and} \\ NaI + C_2H_5OH + C_2H_4 + (CH_3)_2S \end{cases}$$

t = time of reaction in minutes; C_A = concentration of sodium ethoxide (mole litre^{-1}); C_B = concentration of ethyl dimethyl sulphonium iodide (mole litre^{-1}).

t	$C_A \times 10^2$	$C_B \times 10^2$
0	9·625	4·920
12	8·578	3·876
20	8·046	3·342
30	7·485	2·783
42	6·985	2·283
51	6·709	2·005
63	6·386	1·682

From the plot in Fig. 4–2, the rate coefficient, $k = 3.75 \times 10^{-3}$ mole litre^{-1}, s^{-1}.

Suppose $C_S/C_M = 0.1$ at time $t = 0$, then $C_S/C_M = 0.2$ when A is half reacted. Under these conditions, the term in the square brackets in Eqn (4–11) drops from a value of 0.99 at the commencement of the reaction to 0.96 at the half-life. To a first approximation, therefore, this term may usually be assumed constant at an average value, Q, over a fair extent of reaction. With this assumption, Eqn (4–12) becomes the integrated form of Eqn (4–11) with the boundary condition, $C_M = C_{M_0}$ at $t = 0$:

$$kQt = 1/C_M - 1/C_{M_0} \tag{4–12}$$

The rate coefficient in this case may therefore be determined from a plot of $1/C_M$, as ordinate, against t, as abscissa, the graph approximating closely to a straight line with slope equal to kQ.

With some second-order reactions, the stoichiometry is such that species A and B do not react mole for mole. Suppose one mole of A reacts with r moles of B according to the chemical equation,

$$A + rB = \text{products}$$

and the rate law is as given in Eqn (4–13).

$$-dC_A/dt = -dC_B/r \, dt = kC_A C_B \tag{4–13}$$

In this case, Eqn (4–14) represents the integrated form of the rate law.

$$kt = \frac{1}{rC_{A_0} - C_{B_0}} \ln \frac{C_{B_0} C_A}{C_{A_0} C_B} \tag{4–14}$$

An example of a reaction falling into this class is the oxidation of iodide ion to iodine by hydrogen peroxide in aqueous acid solution.

$$H_2O_2 + 2I^- + 2H^+ = I_2 + 2H_2O$$

At a constant pH, this reaction is of the first order in both hydrogen peroxide and iodide ion, following the rate law shown in Eqn (4–15).

$$-dC_{H_2O_2}/dt = -dC_{I^-}/2 \, dt = k^1 C_{H_2O_2} C_{I^-} \tag{4–15}$$

(In this case, the coefficient, k^1, changes with the set concentration of hydrogen ion and is therefore not a simple rate coefficient, but represents a composite value, as discussed on page 78.)

Third-order reactions are not common, but may be treated according to the same principles as those outlined for reactions of simpler order. Suppose, for example, that species A and B are reacting according to the stoichiometric equation,

$$A + rB = \text{products}$$

and that the reaction obeys the third-order rate law given by Eqn (4–16):

$$-dC_A/dt = -dC_B/r\, dt = kC_A^2 C_B \tag{4-16}$$

Applying the boundary conditions, $C_A = C_{A_0}$ and $C_B = C_{B_0}$ at $t = 0$, integration yields the following relationship among the variables:

$$kt = \frac{1/C_A - 1/C_{A_0}}{C_{B_0} - rC_{A_0}} + r\left[\frac{1}{(C_{B_0} - rC_{A_0})^2}\right]\ln\left(\frac{C_A C_{B_0}}{C_{A_0} C_B}\right) \tag{4-17}$$

The rate coefficient may be estimated from a plot of

$$\left(\frac{1}{C_A} - \frac{1}{C_{A_0}}\right) + \left(\frac{2\cdot303r}{C_{B_0} - rC_{A_0}}\right)\log\frac{C_A}{C_B}$$

against t. The reaction between triphenyl methyl chloride and methanol* in benzene solution follows this form of rate law, with $r = 1$.

Complications arise in the use of Eqn (4–17) when $C_{B_0} \approx rC_{A_0}$. As in the case of the mixed second-order reaction, it is convenient to recast the differential equation (4–16) in terms of a concentration, C_i, intermediate between C_B and rC_A such that

$$C_A = C_i - C_s/3r \tag{4-18}$$

and

$$C_B = rC_i + 2C_s/3 \tag{4-19}$$

with,

$$C_s = C_{B_0} - rC_{A_0} \tag{4-20}$$

Recasting the differential equation (4–16),

$$\frac{-dC_i}{dt} = k\left(C_i - \frac{C_s}{3r}\right)^2\left(rC_i + \frac{2C_s}{3}\right) = rkC_i^3\left[1 - \frac{(C_s/rC_i)^2}{3} + 2\frac{(C_s/rC_i)^3}{27}\right] \tag{4-21}$$

For small values of C_s, the term in the square brackets is close to unity, and, to a first approximation, may be assumed constant at an average value, Q, over a fair extent of reaction. Integration with the boundary condition $C_i = C_{i_0}$ at $t = 0$, yields Eqn (4–22) in a convenient form for plotting:

$$2rQkt = 1/C_i^2 - 1/C_{i_0}^2 \tag{4-22}$$

The expressing of a rate coefficient in terms of its correct units requires a knowledge of the complete form of the kinetic law. Owing to the restricted conditions under which a reaction is sometimes studied, a simple rate law may be observed as a limiting case of a more complex law. Thus, the mixed second-order kinetic law given by Eqn (4–5), and involving

---

*Swain, C. G., J. Amer. Chem. Soc., 1948, **70**, 1119.

species A and B, approaches the simple first-order form, $-dC_A/dt =$ constant $\times C_A$, when B is in considerable excess; the rate coefficient for the reaction equals the constant in this simplified equation divided by the concentration of B. Similarly, confusion may arise in the case of an acid-catalysed reaction in which the rate is directly proportional to the concentration of hydrogen ion, the latter remaining unchanged during the course of a particular kinetic run. As an example, the acid-catalysed esterification of acetic acid with ethanol obeys the following rate equation:

$$-dC_{CH_3COOH}/dt = kC_{H^+}C_{CH_3COOH}C_{C_2H_5OH} \tag{4-23}$$

but, for a given kinetic run, the reaction follows second-order kinetics, viz.

$$-dC_{CH_3COOH}/dt = \text{constant} \times C_{CH_3COOH} \times C_{C_2H_5OH}$$

and the hydrogen-ion concentration must be taken into account when evaluating the rate coefficient. A slightly more complicated case is that of the oxidation of iodide ion to iodine by hydrogen peroxide in aqueous acid solution (see page 76). This reaction obeys the rate law given by Eqn (4–24), which, at constant pH, becomes a simple mixed second-order law, Eqn (4–15).

$$-dC_{H_2O_2}/dt = k_1C_{H_2O_2}C_{I^-} + k_2C_{H_2O_2}C_{I^-}C_{H^+} \tag{4-24}$$

The progress of a reaction is often followed by changes in a physical measurement, such as electrical conductivity or the rotation of the plane of polarized light, and the coefficient in the rate equation may be in terms of the physical measurement unless suitable conversion factors to concentration units are allowed for.

4–2   Differential and open-ended methods

The differential form of a kinetic equation is usually much simpler than the integrated form and, given good data, may be used directly for the estimate of a rate coefficient. When a reaction is well behaved, the integrated form yields the more accurate estimate, but it can be cumbersome and difficult to apply in certain cases. The differential form, based on initial rates, is also appropriate to reactions in which the subsequent stages are complicated by undesirable side reactions.

As an example of the use of the differential form of the rate equation, consider its application to the data listed in Table 4–1 for the gas-phase dimerization of 1,3-butadiene. The appropriate differential equation is

$$-dp/dt = 2kp^2 \tag{4-25}$$

the partial pressure of butadiene, p, being related to the total pressure in the system p_T and the initial pressure p_0 according to Eqn (4–26):

$$p = 2p_T - p_0 \qquad (4\text{–}26)$$

The differential equation may be expressed in terms of p_0 and p_T as follows:

$$-dp_T/dt = k(2p_T - p_0)^2$$

$$(-dp_T/dt)^{1/2} = 2k^{1/2}p_T - k^{1/2}p_0 \qquad (4\text{–}27)$$

A graph of $(-dp_T/dt)^{1/2}$, as abscissa, against p_T, as ordinate, should give a straight line with slope equal to $2k^{1/2}$. The method has the advantage of making no assumptions with regard to the magnitude of p_0 (necessary in the case of the integrated form of equation). Values of $-dp/dt$ may be taken as $-\Delta p_T/\Delta t$ for the mid-point of the time interval, Δt, according to Eqn (2–18) (see p. 24), the second-order terms involving $\Delta^2 p_T$ vanishing and the third-order terms involving $\Delta^3 p_T$ being of insignificant magnitude.*

To a close approximation, Eqn (4–27) may be rewritten in the form of Eqn (4–28).

$$(-\Delta p_T)^{1/2} = 2(k\,\Delta t)^{1/2}p_T - (k\,\Delta t)^{1/2}p_0 \qquad (4\text{–}28)$$

The rate coefficient may therefore be estimated from a graph of $(-\Delta p_T)^{1/2}$ against p_T as shown in Fig. 4–3 from the data listed in Table 4–3. From this plot, $k = 1.25 \times 10^{-5}$ mm^{-1} min^{-1} (method of averages) compared with a value of $1.1_4 \times 10^{-5}$ mm min^{-1} from the plot based on the integrated form of the rate equation (Fig. 4–1); in more conventional units the values are 7.8×10^{-3} mole^{-1} l s^{-1} and 7.1×10^{-3} mole^{-1} l s^{-1} respectively. Extrapolation to $\Delta p_T = 0$ in Fig. 4–3 gives an estimated final pressure in the reaction vessel of $p_{T\infty} = 323$ mm; the corresponding value of $P_{T_0}/2$ (Table 4–1) is 317 mm.

E. A. Guggenheim (*Phil. Mag.*, 1926, **2**, 538) has described a method for evaluating the rate coefficient of a first-order reaction which does not require a knowledge of the magnitude of the initial or final concentration of the reacting species. In many instances the course of a reaction may be followed by measurement of some physical property of the system, ϕ, which varies with time according to Eqn (4–29).

$$(\phi_\infty - \phi) = (\phi_\infty - \phi_0)\exp(-kt) \qquad (4\text{–}29)$$

The initial and final values of the physical property are ϕ_0 and ϕ_∞ respectively, and k represents the rate coefficient for a simple, unopposed, first-order reaction. If readings $\phi_1, \phi_2, \ldots, \phi_n$ are made at times $t_1, t_2, \ldots, t_n$,

---

*A small correction may be made for these terms, if necessary.

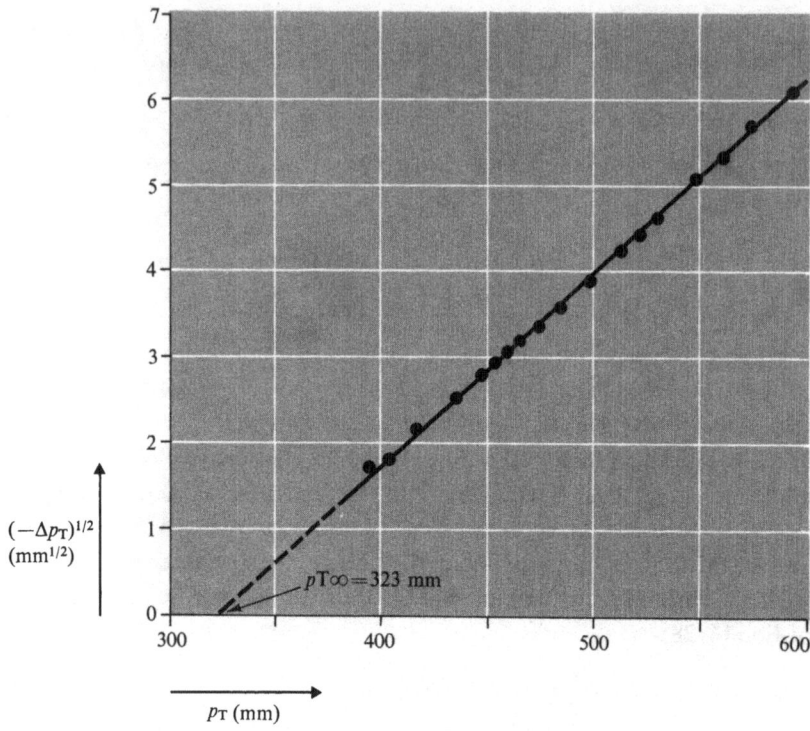

$(-\Delta p_T)^{1/2}$
$(mm^{1/2})$

$pT\infty = 323$ mm

p_T (mm)

Fig. 4-3   Differential plot for the dimerization of 1,3-butadiene (see Table 4-3).

and a second series ϕ'_1, ϕ'_2, ..., ϕ'_n is made at the corresponding times $t_1 + \Delta t$, $t_2 + \Delta t$, ..., $t_n + \Delta t$ (where Δt is constant) then,

$$(\phi_\infty - \phi_n) = (\phi_\infty - \phi_0) \exp(-kt_n) \tag{4-30}$$

and

$$(\phi_\infty - \phi'_n) = (\phi_\infty - \phi_0) \exp[-k(t_n + \Delta t)] \tag{4-31}$$

Subtraction of Eqn (4-31) from Eqn (4-30) yields Eqns (4-32) and (4-33):

$$(\phi'_n - \phi_n) = (\phi_\infty - \phi_0)[1 - \exp(-k\,\Delta t)]\exp(-kt_n) \tag{4-32}$$

$$\ln(\phi'_n - \phi_n) = \ln(\phi_\infty - \phi_0) + \ln[1 - \exp(-k\,\Delta t)] - kt_n \tag{4-33}$$

For a given kinetic run, ϕ_∞, ϕ_0, and Δt are constant, and therefore,

$$\log(\phi'_n - \phi_n) = \text{constant} - kt_n/2 \cdot 303 \tag{4-34}$$

A graph of $\log(\phi' - \phi)$ as abscissa, plotted against time, as ordinate, will be a straight line with slope equal to $-k/2 \cdot 303$. The most appropriate value to be chosen for the time interval, Δt, depends upon the accuracy of the recorded values of ϕ, but for most data Δt should be of the order of one half-life, so that for application of the Guggenheim method the reaction should be well behaved over at least two half-life periods.

Table 4–3   The differential method of estimating the rate coefficient
Dimerization of 1,3-butadiene: see also Table 4–1.

t	p_T	$-\Delta p_T$ (for $t = 10$)	$(-\Delta p_T)^{1/2}$
5	611·0		
10	591·9	37·3	6·11
15	573·7	32·8	5·73
20	559·1	28·0	5·29
25	545·7	25·5	5·05
30	533·6	21·8	4·67
35	523·9	19·9	4·46
40	513·7	18·1*	4·25
50	497·4	14·9	3·86
60	483·9	12·4	3·53
70	472·5	10·8	3·28
80	462·4	9·6	3·09
90	453·4	8·6	2·93
100	445·2	7·6†	2·75
120	432·2	6·2	2·49
140	420·5	5·1‡	2·27
180	404·1	3·4	1·84
220	393·3	2·9	1·70
260	380·9		

* Estimated from $(533·6 - 497·4)/2$.
† Estimated from $(462·4 - 432·2)/4$.
‡ Estimated from $(445·2 - 404·1)/8$.

Figure 4–4 shows a Guggenheim plot based on the data in Table 4–4 for the hydrolysis of diethyl acetal in aqueous acid solution—a reaction which proceeds with a volume change and the progress of which may be conveniently followed with a dilatometer.

$$CH_3CH(OC_2H_5)_2 + H_2O = CH_3CHO + 2C_2H_5OH$$

Related to the Guggenheim plot is a simpler and more versatile plot independently proposed by Kezdy and Swinbourne (F. J. Kezdy, J. Jaz, and A. Bruylants, *Bull. Soc. Chim. Belg.*, 1958, **67**, 687; E. S. Swinbourne, *J. Chem. Soc.*, 1960, 2371). This plot is based upon Eqn (4–35) which is derived from the division of Eqn (4–30) by Eqn (4–31) and subsequent rearrangement.

$$\phi_n = \phi_\infty[1 - \exp(k\,\Delta t)] + \phi_n' \exp(k\,\Delta t) \tag{4–35}$$

The last equation demonstrates that a straight line is obtained when the values of ϕ_n are plotted directly against the corresponding values of ϕ_n': an estimate of the rate coefficient may be evaluated from the logarithm of the slope of this line. Further, for time, $t = \infty$, $\phi_n = \phi_n' = \phi_\infty$; therefore ϕ_∞ is the point on the line at which ϕ_n and ϕ_n' have equal value. The graph may also be used for easy extrapolation to ϕ values outside the range of

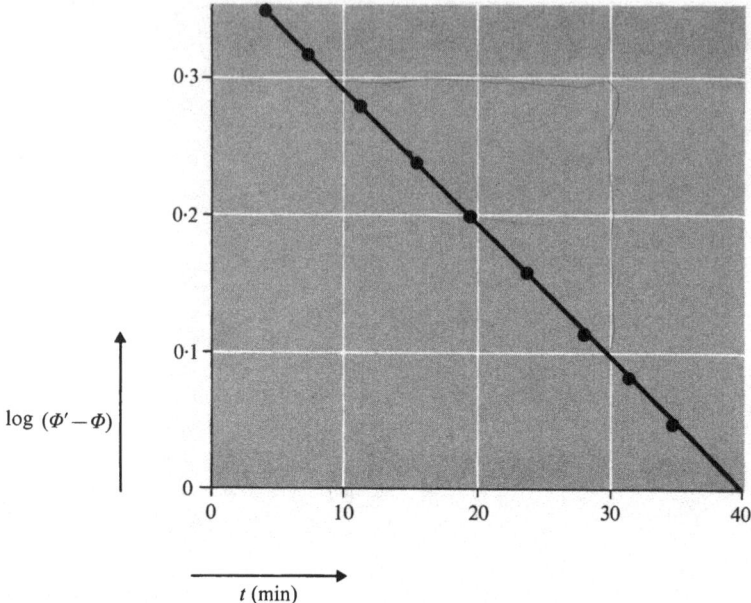

Fig. 4–4 Hydrolysis of diethyl acetal: a first-order plot using the Guggenheim method. Data are from Table 4–4.

Table 4–4 Guggenheim method for first-order reactions

(See Fig. 4–4)

Data are for the hydrolysis of diethyl acetal in aqueous acid solution (pH = 3·5) at 25 °C: a first-order process. t = time of reaction in minutes; ϕ = dilatometer reading in mm (proportional to volume change in the system); ϕ' = dilatometer reading after Δt = 20 min.

t	ϕ	$\phi' - \phi$	$\log(\phi' - \phi)$
4	5·16	2·23	0·3483
8	5·69	2·05	0·3118
12	6·17	1·88	0·2742
16	6·61	1·73	0·2380
20	7·02	1·57	0·1959
24	7·39	1·43	0·1553
28	7·74	1·29	0·1106
32	8·05	1·18	0·0719
36	8·34	1·08	0·0334
40	8·59	1·00	0·0000
44	8·82		
48	9·03		
52	9·23		
56	9·42		
60	9·59		

Rate coefficient $k = 3.83 \times 10^{-4}\ \text{s}^{-1}$ (method of averages).

recorded data. Figure 4–5 demonstrates application of this method to data (Table 4–5) for the gas-phase decomposition of cyclohexyl chloride to cyclohexene and hydrogen chloride at 367·9 °C. This reaction may be followed by measuring the total pressure in the reaction system, p, which varies with time according to Eqn (4–36) (compare with Eqn (4–29)).

$$(p_\infty - p) = (p_\infty - p_0) \exp(-kt) \tag{4–36}$$

Hartley has outlined a method for obtaining, directly from tabulated data for associated variables x and y, a least-squares fit for a straight-line relationship between the differential coefficient, dx/dy, and x (H. O. Hartley, *Biometrika*, 1948, **35**, 32). The procedure, known as the *method of internal least squares*, is based upon finite difference calculus and is clearly applicable to concentration versus time data for a chemical reaction obeying a first-order kinetic law. Hartley's method does not require assumptions concerning the initial and final concentrations of reactant, but it involves somewhat more laborious computations than either the Guggenheim or the Kezdy–Swinbourne methods. Its application to the data in Table 4–5 is illustrated in Example 4–1 at the end of this chapter. Hartley has provided a further extension of the method which is suitable for second-order kinetic systems.

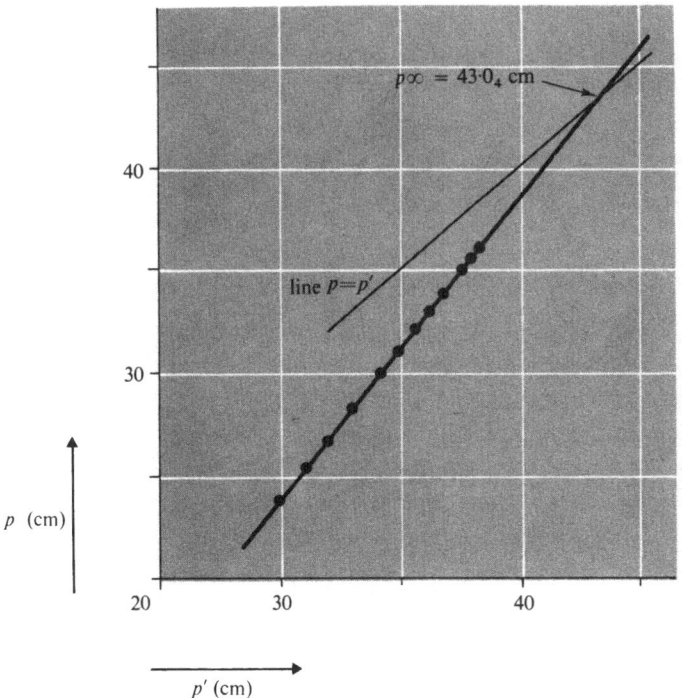

Fig. 4–5 Kezdy–Swinbourne plot for the thermal decomposition of cyclohexyl chloride. Data are from Table 4–5.

Table 4–5   Thermal decomposition of cyclohexyl chloride

(See Fig. 4–5)

Data are for thermal decomposition of gaseous cyclohexyl chloride to cyclohexene
and hydrogen chloride at 368 °C. The decomposition follows the equation,

$$(p_\infty - p) = (p_\infty - p_0)\exp(-kt).$$

t = time in minutes; p = total pressure (cm Hg); p_0 = initial pressure (cm Hg);
p_∞ = pressure at completion of decomposition; p' = total pressure after Δt
= 12 min.

t	p	p
3	23·72	29·90
6	25·53	31·12
9	27·13	32·22
12	28·58	33·21
15	29·90	34·11
18	31·12	34·93
21	32·22	35·69
24	33·21	36·37
27	34·11	36·99
30	34·93	37·55
33	35·69	38·05
36	36·37	38·49
39	36·99	
42	37·55	
45	38·05	
48	38·49	

Slope of graph $= \exp(k\,\Delta t)$ Eqn (4–35)

$\qquad\qquad\quad = 1\cdot467$ (method of averages)

Hence $k = 5\cdot32 \times 10^{-4}\,\mathrm{s}^{-1}$

From graph, $p_\infty = 43\cdot0_4$ cm.

4–3   Opposed, concurrent, and consecutive reactions

There is no standard or straightforward procedure for analysing the
data from a reaction system which does not conform to a simple order or
stoichiometry. Material balance graphs of the type outlined in Chapter 3,
are useful for general classification of such systems, but their analysis in
terms of an interacting pattern of simple reactions becomes an individual
problem for each system, and in some instances it is not possible to
distinguish meaningfully between different interpretations of a given set
of data.

  If the deviation of a system from simple behaviour is slight, it is
often sufficient to assume that the simple relationship still holds, and to
treat the deviation as a correction to it. In other instances a comparison
with relatively simple prototypes of complex systems is of assistance in
categorizing the reaction behaviour and in classifying it in terms of

combinations of simple reaction types. Some of these prototypes are considered in this section.

One simple prototype is that of two opposing reactions, each of the first order:

$$A \underset{k_2}{\overset{k_1}{\rightleftharpoons}} B$$

The net rate of disappearance of A is given by Eqn (4–37):

$$-dC_A/dt = k_1 C_A - k_2 C_B \tag{4-37}$$

Assuming the boundary conditions that at time $t = 0$, $C_A = C_{A_0}$, $C_B = 0$, and that at time $t = \infty$, $C_A = C_{A\infty}$, $C_B = C_{B\infty}$, integration yields Eqn (4–38).

$$(C_A - C_{A\infty}) = (C_{A_0} - C_{A\infty}) \exp\left[-(k_1 + k_2)t\right] \tag{4-38}$$

$C_{A\infty}$, $C_{B\infty}$, k_1, and k_2 are, of course, related through the equilibrium constant, K.

$$C_{B\infty}/C_{A\infty} = k_1/k_2 = K \tag{4-39}$$

Equation (4–38) is analogous to Eqn (4–29) so that the Kezdy–Swinbourne plot (see page 81) may be used to estimate $C_{A\infty}$, $(k_1 + k_2)$ and hence the separate values of k_1 and k_2. (An instance of the use of this method for such a system is given in Problem 4–7 at the end of this chapter.) Substitution for k_2 in Eqn (4–38), leads to Eqn (4–40), a convenient form which involves only k_1.

$$k_1 t = \left|\frac{C_{A_0} - C_{A\infty}}{C_{A_0}}\right| \ln \left|\frac{C_{A_0} - C_{A\infty}}{C_A - C_{A\infty}}\right| \tag{4-40}$$

Two other useful prototypes of opposing reaction systems are: (a) a first-order reaction opposed by a second-order reaction; and (b) two opposing second-order reactions. The first of these may be represented by

$$A \underset{k_2}{\overset{k_1}{\rightleftharpoons}} X + Y$$

The integrated form of the rate equation appropriate to this system is given by Eqn (4–41):

$$k_1 t = \frac{C_{A_0} - C_{A\infty}}{C_{A_0} + C_{A\infty}} \ln \frac{C_{A_0}^2 - C_{A\infty} C_A}{(C_A - C_{A\infty})C_{A_0}} \tag{4-41}$$

If the system is taken as proceeding in the opposite direction, with $C_{X_0} = C_{Y_0}$, and $C_A = 0$ at $t = 0$, then,

$$k_2 t = \left|\frac{C_{X_0} - C_{X\infty}}{2C_{X_0} C_{X\infty} - C_{X\infty}^2}\right| \ln W \tag{4-42}$$

where

$$W = \frac{(C_{X_0} - C_{X\infty})(C_X C_{X_0} + C_{X_0} C_{X\infty} - C_X C_{X\infty})}{C_{X_0}^2 (C_X - C_{X\infty})} \tag{4-43}$$

The last two equations are also applicable to the system,

$$2X \underset{k_1}{\overset{k_2}{\rightleftharpoons}} A$$

with the net rate of disappearance of X given by Eqn (4-44):

$$\frac{-dC_X}{dt} = k_2 C_X^2 - k_1 \frac{C_{X_0} - C_X}{2} \tag{4-44}$$

Two opposing second-order reactions may be represented as follows:

$$A + B \underset{k_2}{\overset{k_1}{\rightleftharpoons}} X + Y$$

If at zero time only A and B are present, at concentrations C_{A_0} and C_{B_0} respectively, then

$$(k_1 - k_2)t = \left(\frac{1}{M}\right) \ln \left| \frac{(C_{A_0} - C_{A\infty})(C_A - C_{A\infty} + M)}{(C_A - C_{A\infty})(C_{A_0} - C_{A\infty} + M)} \right| \tag{4-45}$$

in which,

$$M = [1/K - 1][K^2(C_{B_0} - C_{A_0})^2 + 4C_{A_0} C_{B_0} K]^{1/2} \tag{4-46}$$

K being the equilibrium constant. For equal concentrations of A and B, Eqn (4-47) is applicable:

$$k_1 t = \left| \frac{C_{A_0} - C_{A\infty}}{2C_{A_0} C_{A\infty}} \right| \ln \left| \frac{C_{A_0} C_{A\infty} - 2C_A C_{A\infty} + C_A C_{A_0}}{C_{A_0}(C_A - C_{A\infty})} \right| \tag{4-47}$$

This equation also applies to the system,

$$2A \underset{k_2}{\overset{k_1}{\rightleftharpoons}} X + Y$$

the net rate of disappearance of A being given by Eqn (4-48):

$$-\frac{dC_A}{dt} = k_1 C_A^2 - k_2 \left(\frac{C_{A_0} - C_A}{2}\right)^2 \tag{4-48}$$

The gas-phase decomposition of hydrogen iodide to hydrogen and iodine is an example of this system.

The case of parallel or concurrent reactions provides a further set of prototypes of complex systems. A relatively simple example is that of parallel reactions obeying a common form of kinetic law; for instance, a single substance may be decomposing to three products by three simultaneous processes of the first order:

$$A \begin{cases} \xrightarrow{k_1} X \\ \xrightarrow{k_2} Y \\ \xrightarrow{k_3} Z \end{cases}$$

In this particular instance,

$$-dC_A/dt = k_1 C_A + k_2 C_A + k_3 C_A$$

$$-dC_A/dt = (k_1 + k_2 + k_3)C_A \tag{4-49}$$

$$= k_s C_A \tag{4-50}$$

The overall kinetics are also of the first order.

$$dC_X/dt = k_1 C_A \tag{4-51}$$

and

$$dC_Y/dt = k_2 C_A \tag{4-52}$$

Dividing Eqn (4–51) by Eqn (4–52),

$$dC_X/dC_Y = k_1/k_2 \tag{4-53}$$

Integration of Eqn (4–53), assuming $C_X = C_Y = 0$ at $t = 0$, yields the result,

$$C_X/C_Y = k_1/k_2 \tag{4-54}$$

It follows that,

$$C_X : C_Y : C_Z = k_1 : k_2 : k_3 \tag{4-55}$$

Although Eqn (4–55) has been derived here for processes of the first order, it holds generally for parallel processes obeying a common form of kinetic law. The formation of a mixture of dinitrobenzenes from the reaction of nitrobenzene with nitric acid in sulphuric acid provides an example of a reaction which is of mixed second order, being of the first order in each of nitrobenzene and nitric acid; if in a particular experiment the ratios of the ortho-:meta-:para-isomers of dinitrobenzene were $6.4 : 93.5 : 0.1$, then these would also represent the ratios of the second-order rate coefficients for the reactions producing the isomers.

An interesting example of parallel processes is that of two different reactants producing a common product. Consider the situation when both reactions are of the first order:

$$A \xrightarrow{k_1} X \qquad\qquad B \xrightarrow{k_2} X$$

$$C_X = (C_{A_0} - C_A) + (C_{B_0} - C_B)$$

$$= (C_{A_0} + C_{B_0}) - C_{A_0} \exp(-k_1 t) - C_{B_0} \exp(-k_2 t) \tag{4-56}$$

$$= C_{X_\infty} - C_{A_0} \exp(-k_1 t) - C_{B_0} \exp(-k_2 t) \tag{4-57}$$

therefore

$$\ln(C_{X_\infty} - C_X) = \ln[C_{A_0} \exp(-k_1 t) + C_{B_0} \exp(-k_2 t)] \tag{4-58}$$

If there were only one reactant, or if $k_1 = k_2$, a plot of $\log(C_{X_\infty} - C_X)$ versus t should yield a straight line. If $k_1 > k_2$, and provided C_{A_0} and C_{B_0} are of similar magnitude, the plot of $\log(C_{X_\infty} - C_X)$ versus t is curved for early values of t and approaches a straight line for later values of t, when A has been largely depleted. Under the latter conditions, Eqn (4–58) approaches the form of Eqn (4–59), and C_{B_0} and k_2 may be estimated from the straight-line portion of the graph; with this information, C_{A_0} and k_1 may then be determined as well.

$$\ln(C_{X_\infty} - C_X) \approx \ln C_{B_0} - k_2 t \tag{4-59}$$

H. C. Brown and R. S. Fletcher (*J. Amer. Chem. Soc.*, 1949, **71**, 1845) have used this approach for studying the hydrolysis of a mixture of two isomers (see Fig. 4–6 and Problem 4–10).

Parallel reactions may be of different order. For example, the hydrolysis of an organic halide, RX, may proceed by simultaneous first-order ($S_N 1$) and second-order ($S_N 2$) reactions:

$$-dC_{RX}/dt = k_1 C_{RX} + k_2 C_{RX} C_{OH^-} \tag{4-60}$$

Integration of Eqn (4–60) does not lead to a relationship which is convenient for the estimation of k_1 and k_2 from a set of experimental data. However, rearrangement of Eqn (4–60) to the form Eqn (4–61), shows that k_1 and k_2 may be obtained directly from the plot of $(-dC_{RX}/dt)/C_{RX}$ against C_{OH^-}.

$$(-dC_{RX}/dt)/C_{RX} = k_1 + k_2 C_{OH^-} \tag{4-61}$$

The gas-phase isomerization of 1,1-dimethylcyclopropane (DMC) in the presence of hydrogen chloride* provides another example of parallel first-order and second-order reactions (see Problem 4–8). The isomerization obeys the following rate law:

$$-dC_{DMC}/dt = k_1 C_{DMC} + k_2 C_{DMC} C_{HCl} \tag{4-62}$$

In this instance hydrogen chloride is not consumed during the reaction,

---

*Bullivant, J. and E. S. Swinbourne, *J. Amer. Chem. Soc.*, 1969, **91**, 7703.

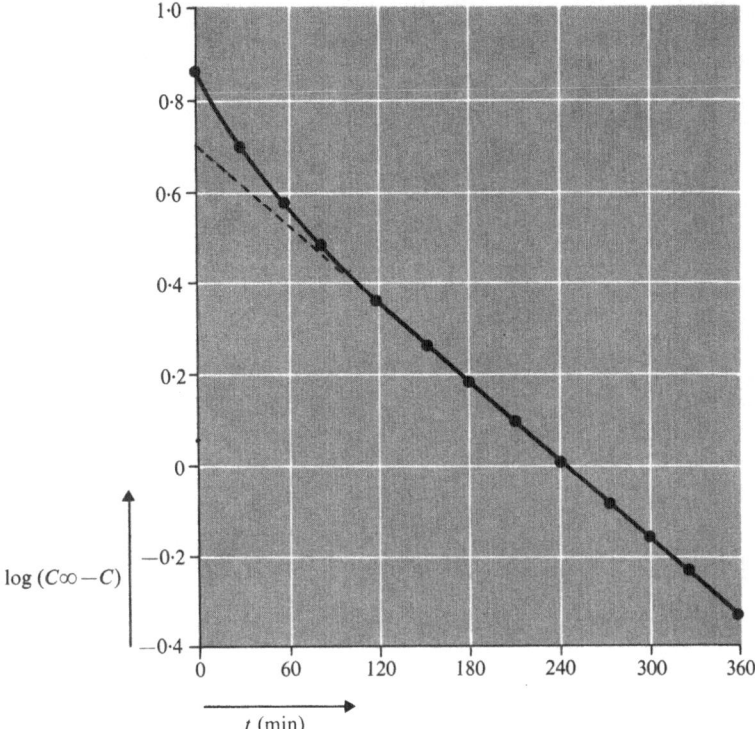

Fig. 4–6 Hydrolysis of a mixture of two aliphatic chlorides. C represents the concentration (arbitrary units) of HCl formed. See Table 4–15 and Problem 4–10 (H. C. Brown and R. S. Fletcher, *J. Amer. Chem. Soc.*, 1949, **71**, 1845).

and so a given kinetic run is of the first order, the overall rate coefficient being equal to $k_1 + k_2 C_{HCl}$. From a number of runs among which the hydrogen chloride concentration is varied, k_1 and k_2 may be estimated by plotting values of the overall rate coefficient against the corresponding values of hydrogen chloride concentration.

With many reaction systems involving free radicals, combination and abstraction reactions of the radicals occur in parallel. For example, when acetone is photolysed, methyl radicals and carbon monoxide are produced, and the main reactions which the methyl radicals undergo are: (a) combination to ethane, and (b) abstraction of hydrogen atoms from acetone to form methane:

$$CH_3COCH_3 \xrightarrow{hv} 2CH_3 + CO \qquad\qquad \text{Step (1)}$$

$$2CH_3 \xrightarrow{k_1} C_2H_6 \qquad\qquad \text{Step (2)}$$

$$CH_3 + CH_3COCH_3 \xrightarrow{k_2} CH_4 + CH_2COCH_3 \qquad\qquad \text{Step (3)}$$

The following rate equations relate to steps (2) and (3):

$$dC_{C_2H_6}/dt = k_1 C_{CH_3}^2 \tag{4-63}$$

$$dC_{CH_4}/dt = k_2 C_{CH_3} C_{CH_3COCH_3} \tag{4-64}$$

The measurement of methyl radical concentration in a system such as this is difficult, but the above rate equations may be combined to give Eqn (4–65), a form which does not involve this concentration term.

$$\frac{dC_{C_2H_6}/dt}{(dC_{CH_4}/dt)^{1/2}} = \frac{k_2}{k_1^{1/2}} C_{CH_3COCH_3} \tag{4-65}$$

If the reaction is conducted under conditions such that the yields of ethane and methane are low, the concentration of acetone remains essentially constant during a kinetic run, and from measurement of this concentration and the relative yields of ethane and methane the value of $k_2/k_1^{1/2}$ may be obtained by application of Eqn (4–65). Independent knowledge of the value of k_1 allows k_2 to be evaluated. Systems similar to this have been exploited considerably in recent years for the study of the rates of methyl radical abstraction reactions and related processes.*

Yet another prototype of complex systems which must be considered is that which involves consecutive reactions (see Fig. 3–5, page 48). Few of the sets of differential equations which define these systems are amenable to analytical integration, but the case of consecutive first-order reactions may be treated exactly. In the latter case the reaction sequence may be represented by

$$A \xrightarrow{k_1} X \xrightarrow{k_2} Y$$

The differential forms of the rate equations are:

$$-dC_A/dt = k_1 C_A \tag{4-66}$$

$$dC_X/dt = k_1 C_A - k_2 C_X \tag{4-67}$$

$$dC_Y/dt = k_2 C_X \tag{4-68}$$

Equation (4–66) leads to the normal first-order decay law,

$$C_A = C_{A_0} \exp(-k_1 t) \tag{4-69}$$

Substitution of the last expression for C_A into Eqn (4–67), and assumption of the boundary condition, $C_X = 0$ at $t = 0$, leads to Eqn (4–70).

$$C_X = C_{A_0}[k_1/(k_2 - k_1)][\exp(-k_1 t) - \exp(-k_2 t)] \tag{4-70}$$

---

* See A. F. Trotman-Dickenson, *Free Radicals*. Methuen, London, 1959.

Further, if $C_Y = 0$ at $t = 0$, then

$$C_A + C_X + C_Y = C_{A_0} \tag{4–71}$$

from which,

$$C_Y = C_{A_0}\left\{1 + \frac{k_2 \exp(-k_1 t) - k_1 \exp(-k_2 t)}{k_1 - k_2}\right\} \tag{4–72}$$

The intermediate, X, rises to a maximum concentration given by Eqn (4–73) at the time of reaction given by Eqn (4–74).

$$C_{X\text{max}} = C_{A_0}\left(\frac{k_1}{k_2}\right)^{k_2/(k_2-k_1)} \tag{4–73}$$

$$t_{\text{max}} = \left(\frac{1}{k_2 - k_1}\right)\ln\left(\frac{k_2}{k_1}\right) \tag{4–74}$$

The maximum might not be observed if some X is present originally in the system.

The rate equations for many complex reaction systems are often more conveniently handled when left in the differential form. Equation (4–61) provides such an example, and a further case is provided by the following consecutive reaction system:

$$A \xrightarrow{k_1} X \quad \text{(first order)}$$

$$A + X \xrightarrow{k_2} Y \quad \text{(mixed second order)}$$

The differential rate equations for the disappearance of A and X are,

$$-dC_A/dt = k_1 C_A + k_2 C_A C_X \tag{4–75}$$

$$-dC_X/dt = -k_1 C_A + k_2 C_A C_X \tag{4–76}$$

Rearrangement of Eqn (4–76) leads to Eqn (4–77); it may be seen that a plot of $-dC_X/C_A\,dt$ against C_X should yield a straight line from which k_1 and k_2 may be determined.

$$-dC_X/C_A\,dt = -k_1 + k_2 C_X \tag{4–77}$$

In this system, it may be observed from Eqn (4–76) that the maximum concentration of X equals k_1/k_2.

It is sometimes useful to eliminate time as an independent variable from a set of differential rate equations. Thus, when Eqn (4–75) is divided by Eqn (4–76) an integrable differential equation results which relates C_A to C_X:

$$\frac{dC_A}{dC_X} = \frac{k_1 + k_2 C_X}{-k_1 + k_2 C_X} \tag{4–78}$$

Using the boundary conditions, $C_A = C_{A_0}$ and $C_X = 0$ at $t = 0$, integration yields Eqn (4–79).

$$C_A = C_{A_0} + C_X + 2\left(\frac{k_1}{k_2}\right) \ln \left(1 - \frac{C_X}{k_1/k_2}\right) \qquad (4\text{–}79)$$

Such equations, of course, yield relative rather than absolute values of the rate coefficients.

The solution of a set of differential equations relating to a complex reaction system is sometimes facilitated by the assumption that the concentrations of certain species (particularly reactive intermediate species) remain essentially constant during the major part of the reaction, thus permitting dC/dt for these species to be equated to zero. This, the *steady-state assumption*, which is described in Chapter 5, has proved most valuable in the treatment of chain reactions.

Although there are many sets of linked differential rate equations for which exact solutions are unavailable, those involving a network of only first-order reactions are solvable. General methods of solution of the latter reaction systems have been outlined by J. Wei and C. D. Prater.[6]

The use of dimensionless parameters, related to time and concentration variables, has been proposed by R. E. Powell for the categorization of both simple and complex reaction systems. The application of this method to simple systems is described in Chapter 3 (see page 56). Further details of the method and its application to complex systems have been given by Frost and Pearson.[1]

4–4 Dead-space corrections

Apparatus used for the study of gas-phase reactions frequently has an appreciable 'dead space' which, although containing reactant, is maintained at a temperature lower than the reaction temperature. For example, the dead space may consist of the pressure gauge and/or the part of the gas-line used for admission of reactant into a static system. When the reaction proceeds with an increase in pressure there is a continual movement of reactant from the reaction zone into the dead space, and when the reaction involves a pressure decrease the movement is in the reverse sense. In either situation, the existence of dead space causes the kinetic behaviour of the system to depart from the 'true' law which would otherwise describe it. Because of the inevitable scatter of experimental readings, the departure may not be readily observable in terms of a non-linearity of the function plot appropriate to the system, but it may nevertheless affect the absolute magnitude of estimate of the rate coefficient—if normal stoichiometry is assumed for the function plot.

For example, in a gaseous decomposition reaction of stoichiometry,

$A = X + Y$, the pressure of reactant, P_r, is given by Eqn (4–80) under ideal conditions.

$$P_r = 2P_0 - P \tag{4–80}$$

where P_0 is the initial pressure in the system and P the total pressure after time of reaction, t.

If the reaction is of the first order, the rate coefficient may be estimated from the slope of a plot of $-\ln (2P_0 - P)$ against t, using the integrated form of the rate equation. The direct use of Eqn (4–80) for a typical reaction system at 400 °C with a 2% dead space at 100 °C can lead to estimates of the rate coefficient which are consistently too low by a factor of about 4%, even though the experimental data may appear to give good straight lines for the function plot. Similar complications arise with other forms of plot (for example, the Guggenheim plot) and different correction factors apply in each case. With very accurate data or with a system having a relatively large dead space the departure from linearity of the function plot may be directly observed.

An equation for the true pressure of reactant present in the reaction zone of a system with dead space has been derived by A. O. Allen[14] assuming that no back-diffusion occurs from the dead space during reaction:

$$P_r = P \frac{q(P_0/P)^a - 1}{q - 1} \tag{4–81}$$

where

$$a = \frac{V_h + (V_c T_h)/T_c}{V_h} \tag{4–82}$$

(q is the stoichiometric number of moles of product formed from one mole of reactant; V_h, T_h are the volume and absolute temperature of the reaction zone; V_c, T_c are the volume and temperature of the dead space). The total pressure in the system at infinite time, P_∞, may be deduced by setting $P_r = 0$ in Eqn (4–81):

$$P_\infty = q^{1/a} P_0 \tag{4–83}$$

For a reaction involving an increase in the total pressure, the linearity of the function plot used for estimating the rate coefficient is improved if Allen's value for P_r (Eqn (4–81)) is used in place of the 'ideal' value such as that given by Eqn (4–80). Alternatively, and more conveniently, P. J. Robinson has shown that even better linearity is obtained if values of P_r as estimated from Eqn (4–84) are used in the function plot.[15]

$$P_r = (P_\infty - P)/(q - 1) \tag{4–84}$$

In the last relationship, either Allen's value for P_∞ may be used (Eqn (4–83)) or, alternatively, an appropriate experimental value, such as that obtained by the Kezdy–Swinbourne method (see page 81). The true rate coefficient is related to the initial slope, S, of the integrated function plot, according to Eqn (4–85), if Robinson's value for P_r is used.

$$k = Sa\left|\frac{P_\infty/P_0 - 1}{q-1}\right|^n \tag{4-85}$$

where n is the order of the reaction. If Allen's value for P_r is used, the corresponding relationship is given by Eqn (4–86).

$$k = S[1/a + q(1 - 1/a)] \tag{4-86}$$

In the case of a gas-phase reaction proceeding with a decrease in total pressure—for example an association reaction—the gas moving from the dead space into the reaction vessel is of constant composition, whereas in the previous case the gas moving in the reverse sense is of variable composition. The relatively simple case of a single reactant undergoing an association reaction is represented stoichiometrically by

$$A = qX$$

with $q < 1$, and the pressure of reactant under ideal conditions (no dead space) is given by Eqn (4–87).

$$P_r = (P - qP_0)/(1 - q) \tag{4-87}$$

When there is dead space, the reactant pressure is given by Eqn (4–88),

$$P_r = [1 + aq(1 - q)]P - [aq(1 - q)]P_0 \tag{4-88}$$

and the pressure at infinite time by Eqn (4–89).

$$P_\infty = \frac{qP_0}{q + (1 - q)/a} \tag{4-89}$$

Equation (4–90) gives the pressure of reactant in terms of P_∞:

$$P_r = [1 + q(a - 1)][P - P_\infty]/[1 - q] \tag{4-90}$$

A linear plot for the integrated form of the rate equation is obtained when the value of P_r is taken from Eqn (4–88) or (4–90) rather than from Eqn (4–87). When either of the two former estimates of P_r are used, the rate coefficient is related to the slopes of the linear plot, according to the following equation:

$$k = Sa/[1 + q(a - 1)] \tag{4-91}$$

It may be noted from Eqn (4–90) that the partial pressure of reactant is directly proportional to $[P - P_\infty]/[1 - q]$. If this latter expression is used in place of P_r in the linear plot, then the rate coefficient is related to the slope of the plot according to Eqn (4–92), where n is the reaction order.

$$k = Sa/[1 + q(a - 1)]^n \qquad (4\text{–}92)$$

Dead-space corrections for more complex gas-phase reaction systems than those described here have been discussed in papers by P. J. Robinson and by A. Maccoll and B. Roberts. Interested readers should consult the References at the end of this chapter.

Examples and problems

Example 4–1

The following is an illustration of the application of the method of internal least-squares (see page 83) to the data given in Table 4–5 for the decomposition of cyclohexyl chloride. The calculations have been made for the first eleven recorded data points only, and the instructions given are for an odd number of data; slightly modified instructions are required for an even number of data.

The first step is the construction of Table 4–6 in which the recorded pressure readings (Table 4–5) are listed in column 3. Zeros are placed at the central positions of columns 1, 2, 4, and 5. The remainder of column 2 consists of the cumulative sums of the p values, summation proceeding towards the centre from both ends of column 3. Second cumulative sums are made in column 1 from the values in column 2. Column 4 consists of special cumulative sums proceeding in both directions from the centre of the table and obtained in the following way: the value 61·02 is obtained by adding the zero in column 4 to the values 31·2 and 29·90 in column 3; the value 119·50 is obtained by adding the 61·02 from column 4 to the values 29·90 and 28·58 from column 3; and so on. Column 5 consists of the squares of the values in column 4. The values in the top halves of columns 2 and 4 are made negative, and the columns are summed. Values in the table are referenced (i), (ii), (iii), (iv), and (v) as shown, and the summations are checked according to the directions given at the base of the table.

The remaining calculations follow the instructions listed in Table 4–7, rounding off of k and p being left to the end of the computations. These values may be compared with those obtained from the Kezdy–Swinbourne method (see page 81).

Problems

4–1 Using the data in Table 4–5, and assuming that the partial pressure of decomposing cyclohexyl chloride is given by $(43\cdot04 - p)$ cm, estimate the rate coefficient for the decomposition reaction using the integrated form of the first-order rate equation.

4–2 Ethyl acetate reacts with sodium hydroxide in aqueous solution to form sodium acetate and ethanol by a mixed second-order process. Experimental data for a particular kinetic run are shown in Table 4–8 in which the concentration of remaining sodium hydroxide is given for a particular time of reaction. Assuming the reaction goes to completion, estimate the rate coefficient using the integrated form of the rate equation.

4–3 Using the differential data listed in Tables 3–2 and 3–8 (see pages 54 and 69) estimate the rate coefficients for the decomposition of gaseous cyclohexyl chloride and acetaldehyde.

4–4 Acetamide reacts with nitrous acid in aqueous solution to produce nitrogen, acetic acid and water according to the following stoichiometric equation:

$$CH_3CONH_2 + HNO_2 = CH_3COOH + N_2 + H_2O$$

Table 4–6 Summation table for the method of internal least-squares

(See data in Table 4–5; further details of the computations are given in Example 4–1 and in Table 4–7)

Col. 1	Col. 2	Col. 3	Col. 4	Col. 5
23·72	−23·72	23·72	−277·12	76 795·49
72·97	−49·25	25·53	−227·87	51 924·74
149·35	−76·38	27·13	−175·21	30 698·54
254·31	−104·96	28·58	−119·50	14 280·25
389·17 (iv)	−134·86 (ii)	29·90	−61·02	3 723·44
0	0	31·12 (i)	0	0
519·14 (v)	170·16 (iii)	32·22	63·34	4011·96
348·98	137·94	33·21	128·77	16 581·71
211·04	104·73	34·11	196·09	38 451·29
106·31	70·62	34·93	265·13	70 293·92
35·69	35·69	35·69	335·75	112 728·06
2110·68	129·97	336·14	128·36	419 489·40

N = no. of p values = 11.

Checks on calculations:

$\sum$ Col. 3 = (i) − (ii) + (iii)

$\sum$ Col. 2 = (v) − (iv)

$\sum$ Col. 4 = N(ii) + N(iii) − 2$\sum$ Col. 2

$\sum$ Col. 1 is checked by a repeat addition.

Table 4–7 Instructions for the method of internal least-squares

(See also Table 4–6 and Example 4–1)

Symbol	Instruction	Value
A	$\sum$ Col. 3	336·14
B	A/N	30·558
C	$\sum$ Col. 4/2	64·18
D	C^2/N	374·4611
E	$\sum$ Col. 5/4	104 872·35
F	$\sum$ Col. 2	129·97
G	$2\sum$ Col. 1 − (iv) − (v)	3 313·05
H	$G − MA/N$	−48·35
L	$E − D + B(G − 2MA/N)$	302·129
M	$N(N^2 − 1)/12$	110
P	$ML − H^2/4$	32 649·7
Q	$−2L/H$	12·498
R	$−FH/2P$	0·096 234
p_∞	$Q + B$	43·06 cm
K	$\log(2 + R) − \log(2 − R)$	0·041 83
k	$(2·303/\Delta t)K$	$5·35 \times 10^{-4}$ s^{-1}

Table 4–8 Reaction of ethyl acetate with sodium hydroxide

(See Problem 4–2)

Time (min.)	C_{NaOH} (moles per litre)
0	0·0266
3	0·0238
6	0·0209
9	0·0190
12	0·0175
15	0·0162
18	0·0151
21	0·0142
24	0·0134
27	0·0128
30	0·0123
∞	0·0064

Table 4–9 Reaction of acetamide with nitrous acid

(See Problem 4–4)

t (min.)	V (ml)
0	0·00
10	5·00
20	8·70
30	11·00
40	12·85
50	14·02
60	14·95
70	15·58

(Reference: F. J. Kezdy, J. Jaz, and A. Bruylants, *Bull. Soc. Chim. Belg.*, 1958, **67**, 687.)

(Problem 4–4 continued)

The production of nitrogen gas closely obeys the following first-order kinetic equation:

$$(V_\infty - V) = (V_\infty - V_0) \exp(-kt) \qquad (4\text{–}93)$$

Table 4–9 lists the volume of nitrogen gas produced at various stages of a kinetic run at 25 °C.

(a) Estimate V_∞, the volume of nitrogen produced after the reaction has gone to completion.

(b) Estimate k, the apparent first-order rate coefficient for the reaction.

4–5 Gaseous vinyl chloride reacts with hydrogen bromide at 372 °C, by a mixed second-order process, to produce vinyl bromide and hydrogen chloride. From the data given in Table 4–10 estimate the rate coefficient for the reaction using the integrated form of the rate equation appropriate to the conditions when the initial concentrations of reactants are nearly equal.

Table 4–10   Reaction of vinyl chloride with hydrogen bromide

(See Problem 4–5)

Temperature, 372 °C. Initial concentration: vinyl chloride $= 12{\cdot}60 \times 10^{-4}$ mole litre^{-1}, hydrogen bromide $= 13{\cdot}20 \times 10^{-4}$ mole litre^{-1}; $t =$ reaction time in minutes; $C =$ concentration of vinyl bromide (mole litre^{-1}).

t	$10^4 \times C$
0	0·00
5	0·12
10	0·27
15	0·40
20	0·54
25	0·65
30	0·78
35	0·90
40	1·01

4–6   Triphenyl methyl chloride reacts with methanol in benzene solution according to the following stoichiometric equation:

$$(C_6H_5)_3CCl + CH_3OH = (C_6H_5)_3COCH_3 + HCl$$

The reaction is of the first order with respect to triphenyl methyl chloride and second order with respect to methanol, when carried out in the presence of pyridine which reduces the significance of the reverse process. Estimate the rate coefficient for this reaction at 25 °C from the data listed in Table 4–11.

Table 4–11   Reaction of triphenyl methyl chloride with methanol

(See Problem 4–6)

Data are for the reaction between triphenyl methyl chloride (initial concentration 0·106 mole litre^{-1}) and methanol (initial concentration 0·054 mole litre^{-1}) in benzene solution at 25 °C. (C. G. Swain, *J. Amer. Chem. Soc.*, 1948, **70**, 1119.) $t =$ reaction time in minutes; $x =$ moles litre^{-1} of triphenyl methyl chloride or methanol which have reacted.

t	x
22	0·0010
174	0·0110
444	0·0207
1510	0·0345
2900	0·0414

4–7   Arsenic acid and iodide ions react in aqueous acid solution to form arsenious acid and triiodide ions, the reaction being reversible:

$$H_3AsO_4 + 3I^- + 2H^+ = H_3AsO_3 + I_3^- + H_2O$$

The rate equations for the forward and back reactions are given by Eqns (4–94) and (4–95) respectively,

$$\text{Forward: } dC_{I_3^-}/dt = k_f C_{H_3AsO_4} C_{I^-} C_{H^+} \tag{4–94}$$

$$\text{Back: } -dC_{I_3^-}/dt = k_b C_{H_3AsO_3} C_{I_3^-}/C_{I^-}^2 C_{H^+} \tag{4–95}$$

If the concentrations of arsenic acid and triiodide ions are kept small compared with those of the other participants, the forward and back reactions may be assumed to be first-order processes apparently only involving changes in the concentrations of arsenic acid and triiodide ions respectively.

Table 4–12 lists data for the forward and back reactions conducted at 60 °C under the pseudo-first-order conditions described above. The reactions were followed by measuring the concentration of I_3^- ions present in terms of the optical density of the solution at a wavelength of 4600 Å: the concentration of I_3^- ions in moles per litre is given by optical density/1010.

(a) Construct Kezdy–Swinbourne plots for the forward and back reactions and estimate the concentrations of triiodide ions at equilibrium.
(b) Estimate the pseudo-first-order rate coefficients for the forward and back reactions, and the corresponding rate coefficients in Eqns (4–94) and (4–95).
(c) Estimate the equilibrium constant for the reaction system at 60 °C.
(D. Britton and Z. Hughes, *J. Chem. Ed.*, 1963, **40**, 607.)

Table 4–12   Reaction of arsenic acid with triiodide ions—forward reaction and reverse reaction at 60 °C.

(See Problem 4–7)

| Time | Optical density | |
(min)	Forward reaction	Reverse reaction
2·5	0·063	0·870
5·0	0·115	0·758
7·5	0·155	0·671
10·0	0·189	0·600
12·5	0·215	0·545
15·0	0·238	0·500
17·5	0·254	0·464
20·0	0·269	0·437
22·5	0·280	0·425
25·0	0·290	0·398

Initial concentrations, forward reaction:

H_3AsO_4	0·00100	mole litre^{-1}
H_3AsO_3	0·0200	mole litre^{-1}
H^+	0·100	mole litre^{-1}
KI	0·400	mole litre^{-1}

Initial concentrations, reverse reaction:

I_3^-	0·00100	mole litre^{-1}
H_3AsO_3	0·0200	mole litre^{-1}
H^+	0·100	mole litre^{-1}
KI	0·400	mole litre^{-1}

4–8   Gaseous 1,1-dimethylcyclopropane isomerizes in the presence of hydrogen chloride at 450 °C to a non-equilibrium mixture of the three isomeric methyl-butenes. The isomerization may be represented by Eqn (4–96), as the sum of an uncatalysed reaction of the first order in dimethylcyclopropane and a catalysed reaction of mixed second order, being first order in each of dimethylcyclopropane (DMC) and hydrogen chloride:

$$-dC_{DMC}/dt = k_1 C_{DMC} + k_2 C_{DMC} C_{HCl} \tag{4–96}$$

For a given kinetic run, in which the concentration of hydrogen chloride remains constant, the overall reaction obeys first-order kinetics, and values of the overall first-order rate coefficient obtained from such runs vary with the pressure of hydrogen chloride according to the data given in Table 4–13. Estimate k_1 and k_2 (units, mole^{-1} litre second^{-1}) from the data in the table.

Table 4–13   Isomerization of 1,1-dimethylcyclopropane in the presence of hydrogen chloride at 450 °C

(See Problem 4–8)

k_{ov} = overall first-order rate coefficient (s^{-1}); p_{HCl} = partial pressure of hydrogen chloride (mm Hg).

p_{HCl}	$10^4 \times k_{ov}$
30	1·9
56	2·8
70	3·0
77	3·0
89	2·9
106	3·2
114	3·9
129	4·7
151	4·9
169	6·1
194	6·3
213	6·4
238	8·0
278	8·7

4–9 The hydrolysis of ethyl acetate to ethanol and acetic acid in the presence of a constant concentration of hydrochloric acid catalyst has been studied by O. Knoblauch (*Z. phys. Chem.*, 1897, **22**, 268).

Table 4–14 lists the concentration of total acid present in a reaction mixture at various reaction times for an original mixture which was 1·000 molar in ethyl acetate, 12·215 molar in water, and 12·215 molar in ethanol. Assuming the forward reaction to be of the first order in each of ethyl acetate and water, and the reverse reaction (esterification) to be of the first order in each of ethanol and acetic acid, estimate the mixed second-order rate coefficients for the forward and reverse reactions.

4–10 A sample of diethyl-t-butyl carbinyl chloride consists of two isomers A and B, which undergo hydrolysis at different rates in alcohol-water mixtures at 25 °C. Each reaction apparently proceeds to completion and is of the first order in the particular isomer. Table 4–15 lists data obtained by H. C. Brown and R. S. Fletcher (*J. Amer. Chem. Soc.*, 1949, **71**, 1845) for a study of the hydrolysis, the amount of hydrochloric acid formed being estimated by titration with standard alkali. From the data construct a graph similar to that shown in Fig. 4–6, and estimate the rate coefficient for the slower hydrolysis (isomer B) according to the method outlined on page 88. From the graph estimate the proportions of isomers A and B originally present, and hence the rate coefficient for the faster hydrolysis (isomer A). (*Note:* As the reactions are each of the first order, titration figures may be used directly in the calculations in place of concentrations in standard units.)

Table 4–14 Hydrolysis of ethyl acetate

(See Problem 4–9)

t = time of reaction in minutes; x = titration figure in millilitres of 0·0612 normal $Ba(OH)_2$ solution for a 1 ml sample of reaction mixture.

t	x
0	7·68
78	8·95
94	9·20
138	9·72
169	9·99
348	11·10
415	11·35
464	11·48
∞	12·00

Table 4–15 Hydrolysis of a mixture of isomers

(See Problem 4–10)

t = time of reaction in minutes; x = titration figure in millilitres of NaOH; solution for a 5 ml sample of reaction mixture.

t	x
0	0
30	2·24
60	3·54
90	4·37
120	4·96
150	5·39
180	5·75
210	6·02
240	6·22
270	(6·41)
300	6·56
330	(6·68)
360	6·78
∞	7·25

4–11 The data given in Table 4–16 are for a reaction system in which a substance, A, decomposes to substance X, which in turn decomposes to substance Y, both reaction steps being apparently irreversible.

(a) Construct a diagram showing the concentrations of the three species at different times of reaction.

(b) By means of a suitable differential plot determine whether the data most closely represents,

(i) two consecutive first-order reactions; or

(ii) a first-order reaction followed by one of second order; or

(iii) a second-order reaction followed by one of first order; or

(iv) two consecutive second-order reactions.

(c) Estimate the rate coefficient for each of the appropriate reaction steps identified in (b).

Table 4–16 Consecutive reactions

(See Problem 4–11)

t = time of reaction in minutes; C = concentration in mole litre^{-1}; reaction scheme:

$A \rightarrow X \rightarrow Y.$

t	C_A	C_X
0	1·000	0·000
5	0·905	0·089
10	0·819	0·159
15	0·741	0·212
20	0·670	0·253
25	0·606	0·282
30	0·549	0·301
35	0·497	0·314
40	0·449	0·320
45	0·407	0·321
50	0·368	0·318
55	0·333	0·312
60	0·301	0·304
65	0·272	0·293
70	0·247	0·282
75	0·223	0·269
80	0·202	0·257
85	0·183	0·243
90	0·165	0·230
95	0·150	0·217
100	0·135	0·203
105	0·122	0·191
110	0·111	0·178
115	0·100	0·167
120	0·091	0·122

4–12 A gaseous decomposition reaction of the first order and of stoichiometry, $A = X + Y$, is proceeding in a static system which has 2·5% dead space. Table 4–17 lists experimental data for such a reaction conducted in a vessel at 385 °C, the dead space being at 120 °C.

(a) Estimate the partial pressure of reactant in the vessel after a reaction time of 10 min,
   (i) assuming zero dead space; and
   (ii) assuming Allen's equation (Eqn (4–81)).

(b) Estimate the total pressure in the vessel at infinite time,
   (i) assuming Allen's equation (Eqn (4–83)); and
   (ii) by the Kezdy–Swinbourne method (see page 81).

(c) Assuming the reaction to be irreversible, estimate the rate coefficient by means of the integrated form of the rate equation,
   (i) assuming zero dead space (Eqn (4–80)); and
   (ii) using Robinson's method (Eqns (4–84) and (4–85)).

Table 4-17   Gaseous decomposition in a system with dead space

(See Problem 4-12)

t = time of reaction in minutes; p = total pressure in reaction vessel (cm Hg).

t	p
0	15·06
2	17·62
4	19·70
6	21·40
8	22·81
10	23·98
12	24·93
14	25·73
16	26·37
18	26·90
20	27·34
22	27·70
24	27·99

References

General references

1. Frost, A. A., and R. G. Pearson. *Kinetics and Mechanism.* John Wiley, New York, 2nd ed., 1961. See particularly Chapters 3 and 8.
2. Laidler, K. J. *Chemical Kinetics.* McGraw-Hill, New York, 2nd ed., 1965. See particularly Chapter 1.
3. Benson, S. W. *The Foundations of Chemical Kinetics.* McGraw-Hill, New York, 1960. See particularly Chapters 2 and 3.
4. Friess, S. L., E. S. Lewis, and A. Weissberger (Eds.). *Investigation of Rates and Mechanisms of Reactions* (vol. 8 of *Technique of Organic Chemistry*). John Wiley (Interscience), New York, 2nd ed., 1961. See particularly Chapter 5 (by R. Livingston) and Chapter 8 (by G. A. Russell) in Part 1.
5. Boudart, M. *Kinetics of Chemical Processes.* Prentice-Hall, Englewood Cliffs, N.J., 1968. See particularly Chapter 10.
6. Wei, J., and C. D. Prater. The structure and analysis of complex reaction systems. *Advances in Catalysis* (D. D. Eley, P. W. Selwood, and P. B. Weisz (Eds.)). Academic Press, New York, 1962, **13**, 203.
7. Rabinovitch, B. The Monte Carlo method. Plotting the course of complex reactions. *J. Chem. Ed.,* 1969, **46**, 262.

References to computer methods

8. Dickson, T. R. *The Computer and Chemistry.* W. H. Freeman, San Francisco, 1968.
9. Wiberg, K. B. *Computer Programming for Chemists.* Benjamin, New York, 1965.
10. De Tar, D. F. (Ed.). *Computer Programs for Chemistry* (3 vols.). Benjamin, New York, 1969.
11. Crossley, T. R., and M. A. Slifkin. Solution of chemical kinetic problems and simulation of kinetics of chemical interactions by analogue computer. *Education in Chemistry,* 1967, **4**, 280.
12. Griswold, R., and J. F. Haugh. Analogue computer simulation. An experiment in chemical kinetics. *J. Chem. Ed.,* 1968, **45**, 576.
13. Higgins, J. *Investigation of Rates and Mechanisms of Reactions,* Chapter 7 (see reference 4, above).

References to dead-space corrections

14. Allen, A. O. *J. Amer. Chem. Soc.*, 1934, **56**, 2053.
15. Robinson, P. J. *Trans. Faraday Soc.*, 1965, **61**, 1655, and 1967, **63**, 2668.
16. Maccoll, A., and B. Roberts. *Trans. Faraday Soc.*, 1966, **62**, 1169.

5 Special considerations

With certain kinetic systems it may be reasonably assumed that the concentrations of some or all of the participating species do not change appreciably with time. This assumption of time-invariance may be applied, for example, to all species for a volume element in a fixed position within a flow system, or to particular species, such as reactive intermediates in a static system. In such cases, the concentrations of these species are maintained in a dynamic balance of *steady state* by the interaction of opposing processes which may be either physical or chemical in origin: bulk flow of material, diffusion, and precipitation constitute typical physical processes, while chemical processes may originate from thermal, photochemical, or electrochemical sources. In this chapter two important types of steady-state systems are discussed. These are the flow system, in which the physical process of flow is opposed by the chemical process of reaction, and the radical-chain system in which the opposing acts of formation and removal of reactive intermediates occur by chemical processes.

5–1 Tubular flow-reactors

Although static systems have been traditionally favoured for accurate kinetic studies, flow systems offer particular advantages for the study of fast reactions or for processes in which a low degree of conversion is required. Such systems are also widely used in industrial practice where large quantities of products may be obtained with limited reactor space; accordingly, analysis of these systems has been extensively developed in the chemical engineering literature (see, for example, the References at the end of this chapter).

The most common flow system is the tubular flow-reactor. In its simplest form this consists of a long tube which is maintained at the required reaction temperature, and through which there is a steady passage of reagent material. There is little longitudinal mixing of material within this type of reactor, and the degree of conversion to products therefore varies in a differential manner from entrance to exit.

The elementary model of flow commonly assumed for the tubular reactor is one of *plug flow*, also called piston flow. This model assumes that, for reagent material, cylindrical cross-sectional volume elements behave as if they were enclosed between pairs of piston faces moving along the reactor. A further assumption made is that, within such elements of infinitesimal thickness, the composition, temperature, and fluid properties of the reagent material are uniform.

There is also negligible diffusion relative to the fluid flow. Under steady-state conditions, all volume elements of reagent material would have identical histories of change as they moved through the reactor: they would have identical residence times within the reactor and undergo the same sequential changes in composition, pressure, and temperature.

A further simplifying assumption inherent in the derivation of the basic kinetic equation for a tubular flow-reactor is one of constancy of the volume of the element of reagent material as it passes through the reactor. Volume changes may result from the existence of a significant pressure differential along the reactor, from the chemical change itself, or from both sources. The assumption of constant volume is reasonable when the pressure differential is small and when the degree of conversion is low; moderate degrees of conversion may be accommodated when the reactant material is adequately diluted with a 'carrier'. The effect of chemical change upon the volume element is particularly important to recognize for gas reactions which proceed with a change in the total number of molecules: this matter is examined in the next section of the chapter.

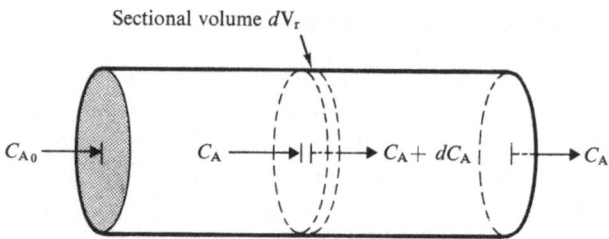

Sectional volume dV_r

C_{A0} C_A $C_A + dC_A$ C_{A1}

Fig. 5–1 Flow of reagent material through a tubular flow-reactor.

Figure 5–1 illustrates the flow of reagent material through a cylindrical sectional volume, dV_r, of infinitesimal thickness in a tubular flow-reactor of total volume, V_r, operating isothermally under idealized plug-flow conditions. For a given time span, dt, the net increase in the number of moles of a given reactant within the sectional volume equals the number of moles flowing into the volume minus the number of moles flowing out of the volume, and minus also the number of moles disappearing as a result of chemical reaction. Equation (5–1) shows this relationship for reactant, A, disappearing chemically according to the conventional rate law, $-dC_A/dt = kC_A^n$, when the volume rate of flow of reagent material is u, the concentration of reactant entering the volume is C_A, and the concentration leaving is $C_A + dC_A$.

$$dN_A = uC_A\,dt - u(C_A + dC_A)\,dt - kC_A^n\,dV_r\,dt \tag{5–1}$$

Rearrangement leads to Eqn (5–2):

$$dN_A/dt = -u\,dC_A - kC_A^n\,dV_r \tag{5–2}$$

Under steady-state conditions, $dN_A/dt = 0$, as the number of moles of A within the sectional volume remains dynamically constant. For such conditions, and with the rate of flow held constant, Eqn (5–3) therefore applies.

$$-dC_A/d(V_r/u) = kC_A^n \tag{5–3}$$

This equation is of the same form as the conventional rate law, $-dC_A/dt = kC_A^n$, with the parameter (V_r/u) replacing the time parameter, t.

The usefulness of Eqn (5–3) is obvious, as it allows for such a flow system, the direct adoption of the forms of equation derived for static systems. Thus, for a reactant disappearing chemically according to a first-order law, Eqn (5–4) may be applied when the concentration of reactant upon entry to the tubular reactor is C_{A_0} and upon exit from the reactor is C_{A_1} (compare with Eqn (4–3)

$$k(V_r/u) = \ln C_{A_0} - \ln C_{A_1} \tag{5–4}$$

Equation (5–4) may be rearranged into the form of Eqn (5–5) (compare with Eqn (4–4) on page 72), from which it may be observed that, for a number of separate kinetic runs conducted with differing flow velocities and a common value of C_{A_0}, values of $\log C_{A_1}$ (abscissa) should plot as a straight line against the corresponding values of V_r/u (ordinate); the slope of the line is equal to *minus* $k/2{\cdot}303$.

$$\log C_{A_1} = \log C_{A_0} - (k/2{\cdot}303)(V_r/u) \tag{5–5}$$

The parameter, V_r/u, is called the *space time*. In the simple model just discussed it equals the residence time for each volume element of reagent material passing through the tubular reactor, but in more complex cases this simple analogy is not maintained.

The plug-flow assumption is reasonable for well-designed reactors, particularly when gaseous reagents are involved. A comparison of this with other forms of flow in tubular reactors has been made by Denbigh (see Fig. 5–2) who has discussed additional sources of error in the use of the ideal reactor model.[1] A brief summary of errors has also been given by Benson (S. W. Benson, *The Foundations of Chemical Kinetics*, McGraw-Hill, New York, 1960, pp. 61–3).

Substantial errors arising from the occurrence of laminar flows are more likely with liquid than gaseous reagents, and this type of flow is enhanced in long, straight, cylindrical reactors. Pressure drops are likely to be significant in long reactors, especially when high flow rates are employed

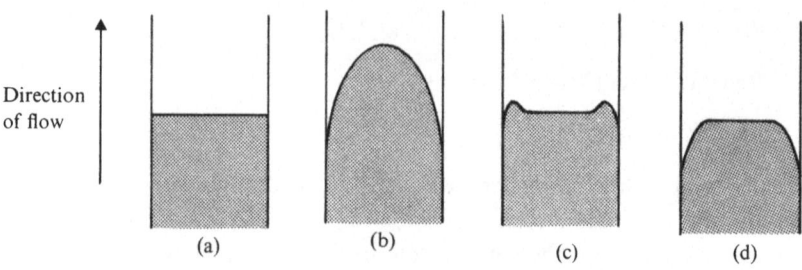

Fig. 5-2   Typical flow profiles in a tubular flow-reactor. (a) Plug flow. (b) Laminar flow. (c) Flow through packing. (d) Turbulent flow. (After Denbigh.[1])

or when the reactor is packed: allowances may be made by assuming an average pressure over small increments of reactor length. A major difficulty in many flow systems is the accurate designation of temperature for the reacting medium. This and the associated difficulty of defining a precise reaction time constitute the two major sources of error in the use of tubular flow-reactors for the provision of accurate kinetic data.

5-2   Reaction systems with variable volume

The convention of expressing reaction rate in terms of the change in concentration of reactant species per unit time (see p. 50) is inappropriate for systems with variable volume. For these systems, relationships which involve the total number of moles of reactant species are preferable. Equations (5–6), (5–7), and (5–8) are examples of such relationships for first-order and second-order cases.

(i) First order:

$$-dN_A/dt = kN_A \tag{5–6}$$

(ii) Second order, one species:

$$-dN_A/dt = kN_A^2/V \tag{5–7}$$

(iii) Second order, two species:

$$-dN_A/dt = kN_A N_B/V \tag{5–8}$$

The last three equations may be expressed in the alternative forms given by Eqns (5–9) to (5–14) which may be compared with the corresponding conventional forms involving change in concentration with time (see section 3–2). Equations (5–9) to (5–14) only become identical with the conventional forms when V is held constant.

(i) $$(1/V)(-dN_A/dt) = kN_A/V \tag{5–9}$$

$$= kC_A \tag{5–10}$$

(ii) $(1/V)(-dN_A/dt) = kN_A^2/V^2$ (5–11)

$$= kC_A^2$$ (5–12)

(iii) $(1/V)(-dN_A/dt) = kN_A N_B/V^2$ (5–13)

$$= kC_A C_B$$ (5–14)

As an example of a closed reaction system with variable volume, consider a first-order gas-phase reaction of the type,

$$A \rightarrow bB$$

proceeding at constant temperature and pressure. Assuming ideal gas behaviour, the ratio of the volume of the system to the initial volume, V/V_0, is given by Eqn (5–15), when the ratio of the remaining number of moles of A to the initial number present is N_A/N_{A_0}.

$$V/V_0 = b + (1-b)N_A/N_{A_0}$$ (5–15)

Integration of Eqn (5–6) gives Eqn (5–16) which applies for reaction time, t.

$$N_A/N_{A_0} = \exp(-kt)$$ (5–16)

Substitution of Eqn (5–16) into Eqn (5–15) yields Eqns (5–17) to (5–19) which give the dependence of V/V_0 upon the time of reaction.

$$V/V_0 = b + (1-b)\exp(-kt)$$ (5–17)

$$\log(V/V_0 - b) = -kt/2{\cdot}303 + \log(1-b) \quad (\text{form for } b < 1)$$ (5–18)

$$\log(b - V/V_0) = -kt/2{\cdot}303 + \log(b-1) \quad (\text{form for } b > 1)$$ (5–19)

If the reaction is being followed by the change in total volume of the system, it is clear from Eqns (5–18) and (5–19) that a plot of $\log(V/V_0 - b)$ or $\log(b - V/V_0)$, as abscissa, against t, as ordinate, yields a straight line with slope equal to *minus* $k/2{\cdot}303$.

The dependence of concentration upon time of reaction in the case just considered is given by Eqns (5–20) and (5–21) which are derived from the division of Eqn (5–17) by Eqn (5–16).

$$C_{A_0}/C_A = b\exp(kt) + 1 - b$$ (5–20)

$$\log(C_{A_0}/C_A - 1 + b) = kt/2{\cdot}303 + \log b$$ (5–21)

Thus, a straight-line graph with slope equal to $k/2{\cdot}303$ is obtained from the plot of $\log(C_{A_0}/C_A - 1 + b)$, as abscissa, against t, as ordinate. When the reaction is conducted under conditions of constant volume, the same slope of graph is, of course, obtained from the plot of $\log(C_{A_0}/C_A)$ against t (see Eqn (4–4)).

The need to use special forms of the rate equations for systems with variable volume has been stressed by a number of workers* and is of particular importance in flow systems involving gaseous reagents.

For example if the gas-phase reaction of the type,

$$A \rightarrow bB \quad (b \neq 1)$$

is being conducted in a tubular flow-reactor operating under conditions of constant temperature and pressure, the volume of the element of reagent material alters continuously as it passes through the reactor. Thus, if $b > 1$, there is an increase in volume as chemical change takes place, and, as a result, the volume rate of flow increases; dilution of reactant occurs, not only because of chemical change but also because of the associated increasing volume of the reagent element. Because the *volume rate of flow* may vary from point to point along a tubular reactor, many workers prefer to use *mass rate of flow*, as this does not alter along the reactor.

Equation (5–22) is the general relationship of the total volume of the reactor, V_r, to the volume rate of flow, u, and residence time, t, of each infinitesimal reagent element, assuming plug flow.

$$V_r = \int_0^t u \, dt \tag{5–22}$$

With plug flow, each infinitesimal volume element of reagent material behaves as a closed reaction system as it moves along the reactor. (This is, of course, the reason for the resemblance of Eqn (5–3), for a flow system, to the corresponding rate equation for a static system.) It follows, therefore, that, in the case of the gas-phase reaction system just discussed, Eqn (5–15) may be adopted to relate u to the initial rate of flow, u_0, when N_{A_0} now corresponds to the initial number of moles of A entering the reactor per unit time, and N_A to the number of moles passing a given point in the reactor per unit time. The required relationship is given by Eqn (5–23).

$$u = u_0[b + (1-b)N_A/N_{A_0}] \tag{5–23}$$

Substitution in Eqn (5–22) leads to Eqns (5–24) and (5–25). (N_{A_1} is the number of moles of A leaving the reactor per unit time, after total residence time, t, of each volume element.)

$$V_r = u_0 \int_0^t [b + (1-b)N_A/N_{A_0}] \, dt \tag{5–24}$$

---

*See, for example, A. F. Benton, *J. Amer. Chem. Soc.*, 1931, **53**, 2984; G. M. Harris, *J. Phys. Chem.*, 1947, **51**, 505; K. G. Denbigh, *Principles of Chemical Equilibrium.* Cambridge University Press, London, 1955, p. 439; M. S. Peters and E. J. Skorpinski, *J. Chem. Ed.*, 1965, **42**, 329.

$$V_r = u_0 \int_{N_{A_1}}^{N_{A_0}} \left\{ \frac{b + (1-b)N_A/N_{A_0}}{-dN_A/dt} \right\} dN_A \tag{5–25}$$

The kinetic behaviour of each volume element, as if it were a closed reaction system, permits the substitution of rate expressions such as those in Eqns (5–6) and (5–7)* for the term $-dN_A/dt$ in Eqn (5–25).

The resultant integrated relationships for the first-order and second-order cases (one reactant species) are given by Eqns (5–26) and (5–27); these may be used to estimate the rate coefficient from experimental data.
First order:

$$k(V_r/u_0) = b \ln (N_{A_0}/N_{A_1}) + (1-b)(1 - N_{A_1}/N_{A_0}) \tag{5–26}$$

Second order:

$$k(V_r N_{A_0}/u_0^2) = b^2(N_{A_0}/N_{A_1} - 1) + 2b(1-b) \ln (N_{A_0}/N_{A_1})$$
$$+ (1-b)^2(1 - N_{A_1}/N_{A_0}) \tag{5–27}$$

When b is made equal to unity, the first term only remains on the right-hand side of each of the last two equations: this would correspond to the situation in which the volume of the reagent material remains constant during its passage through the reactor.

5–3   Stirred flow-reactors

Because of the difficulties associated with the use of tubular flow-reactors for the provision of accurate kinetic data, increasing attention is being given to stirred-flow methods. These have been adapted to both liquid-phase and gas-phase systems. Particular reference should be made to a review by W. C. Herndon[5] and to the papers by K. G. Denbigh and others [1,6] in which the basic theory has been developed.

With a stirred flow-reactor, the composition of material within the whole reaction vessel is kept nearly uniform by means of vigorous mixing; this also ensures a uniformity of temperature and pressure within the vessel. Reagent material flowing into the reactor is distributed almost immediately throughout the vessel, and, ideally, the composition of the effluent is identical with that of the contents of the vessel.

The form of steady-state equation derived for the tubular flow-reactor (see Eqns (5–1)–(5–3)) may be modified for the stirred flow-reactor by replacing the infinitesimal segment of reactor volume considered in the former case by the total volume of the reactor, V_r, and the differential concentration of reactant across the segment by the finite difference in the

*The term V in Eqn (5–7) corresponds to u in the present case.

concentrations of reactant material entering and leaving the stirred flow-reactor $(C_{A_0} - C_{A_1})$. Equation (5–28) is the correspondingly modified form of Eqn (5–3).

$$Rate\ of\ reaction = kC_{A_1}^n = (u/V_r)(C_{A_0} - C_{A_1}) \tag{5–28}$$

(As before, u is the volume rate of flow of reagent material.) If the reaction is of the first order, Eqn (5–29) applies.

$$k = (u/V_r)(C_{A_0} - C_{A_1})/C_{A_1} \tag{5–29}$$

Equation (5–30) gives the corresponding expression for the rate coefficient in the case of two species, A and B, reacting according to the simple second-order kinetic law, $-dC_A/dt = kC_A C_B$.

$$k = (u/V_r)(C_{A_0} - C_{A_1})/C_{A_1} C_{B_1} \tag{5–30}$$

(C_{A_1} and C_{B_1} are the concentrations of A and B in the vessel.)

An important feature of the stirred-flow method is that expressions for the rate coefficient, such as those given by Eqns (5–29) and (5–30), are of simple algebraic forms. This characteristic facilitates the analysis of experimental data from complex reaction systems. As an example, consider the case of a reaction system which is believed to consist of two simultaneous first-order processes of the type,

$$A \xrightarrow{k_1} B$$

$$A \xrightarrow{k_2} C$$

Application of Eqn (5–29) leads to Eqns (5–31) and (5–32).

$$k_1 = (u/V_r)(C_{B_1}/C_{A_1}) \tag{5–31}$$

$$k_2 = (u/V_r)(C_{C_1}/C_{A_1}) \tag{5–32}$$

If the mechanism is as postulated, values of C_{B_1}/C_{A_1} and C_{C_1}/C_{A_1}, obtained from a number of experiments in which u and C_{A_0} are varied, should lie on a pair of straight lines passing through the origin, when plotted against the corresponding values of V_r/u. The rate coefficients, k_1 and k_2 may be estimated directly from the slopes of the straight lines.

The case of two consecutive reactions of the first order,

$$A \xrightarrow{k_1} B \xrightarrow{k_2} C$$

may also be handled conveniently by stirred-flow methods. Equation (5–28) may be expressed in the form of Eqn (5–33) for this particular case.

$$Rate\ of\ reaction\ for\ B = k_2 C_{B_1} - k_1 C_{A_1}$$

$$k_2 C_{B_1} - k_1 C_{A_1} = (u/V_r)(C_{B_0} - C_{B_1}) \tag{5–33}$$

For $C_{B_0} = 0$, Eqn (5–34) applies.

$$u/V_r = k_1 C_{A_1}/C_{B_1} - k_2 \qquad (5\text{–}34)$$

Values of u/V_r (abscissa) should plot as a straight line against values of C_{A_1}/C_{B_1} (ordinate), the slope of the graph being equal to k_1 and the intercept, *minus* k_2.

Further examples of the use of the stirred flow-reactor for the analysis of complex reaction systems have been given by W. C. Herndon[5] and by L. P. Hammett and H. H. Young (*J. Amer. Chem. Soc.*, 1950, **72**, 280).

Equations (5–28) to (5–34) apply to systems in which there is no volume change resulting from chemical reaction. Modifications of the kinetic equations, to allow for volume changes of this nature, may be made along the same lines as those dicussed in section 5–2. Thus, in the case of a first-order gas-reaction of the type, $A \rightarrow bB$, Eqn (5–35) may be used for estimating the rate coefficient when pure A is fed into the stirred flow-reactor (ideal gas behaviour is assumed).

$$k = [u_0 N_{B_1}/V_r(bN_{A_0} - N_{B_1})][1 + (b-1)N_{B_1}/bN_{A_0}] \qquad (5\text{–}35)$$

(N_{A_0} represents the number of moles of A entering the reactor per unit time, and N_{B_1} represents the number of moles of B leaving per unit time; u_0 is the volume rate of flow of reagent material into the reactor.)

When a carrier gas is being used, the form given by Eqn (5–36) is convenient.[8]

$$k = \left[\frac{N_{B_1} R T}{b P V_r}\right] \left[\frac{b(N_{A_0} + N_c) + (b-1)N_{B_1}}{bN_{A_0} - N_{B_1}}\right] \qquad (5\text{–}36)$$

(N_c is the number of moles of carrier gas flowing per unit time; R is the ideal gas constant; T is the absolute temperature; P is the total pressure in the reactor; other terms are as in Eqn (5–35).)

Further equations for these and related systems have been given by G. M. Harris (*J. Phys. Chem.*, 1947, **51**, 505) and by E. S. Lewis and W. C. Herndon (*J. Amer. Chem. Soc.*, 1961, **83**, 1955). The latter authors have pointed out that the forms of the kinetic equations are often simplified by considering the rate of flow of the effluent gas rather than that of the gas entering the reaction vessel.[5]

5–4   Chain reactions

The kinetic equations relating to many complex reaction systems may be simplified by assuming that the concentrations of certain intermediate species remain in approximately steady-state conditions compared with those of other species in the system. The assumption is particularly useful in the case of chain reactions, especially those involving radical species which are very reactive, and therefore present in very small concentrations.

As an example of the application of the steady-state principle to a radical-chain reaction, the thermal decomposition of 1,2-dichloroethane will be considered; this reaction has been studied by D. H. R. Barton and K. E. Howlett (*J. Chem. Soc.*, 1949, 155; *Trans. Faraday Soc.*, 1952, **48**, 25) using a static system.

Near 400 °C 1,2-dichloroethane pyrolyses to vinyl chloride and hydrogen chloride. The reaction is characterized by marked induction periods (see Fig. 3–4) and is very sensitive to the presence of small amounts of olefinic substances, such as propylene, which have an inhibitory effect; these features are characteristic of a radical-chain process. After the induction period, decomposition proceeds smoothly and is of the first order. Arrhenius parameters reported by Barton and Howlett are $E = 47000$ cal mole^{-1} and $\log A = 10\cdot81$; alternative values of $E = 53100$ cal mole^{-1} and $\log A = 13\cdot35$ have been reported by P. Goldfinger *et al.* in *Chem. Rev.*, 1963, **63**, 355.

It has been proposed by Howlett that the radical-chain decomposition of 1,2-dichloroethane proceeds by the following dominant steps:

initiation

$$ClCH_2CH_2Cl \xrightarrow{k_1} ClCH_2CH_2 + Cl \qquad \text{Step (1)}$$

propagation

$$Cl + ClCH_2CH_2Cl \xrightarrow{k_2} ClCH_2CHCl + HCl \qquad \text{Step (2)}$$

$$ClCH_2CHCl \xrightarrow{k_3} CH_2CHCl + Cl \qquad \text{Step (3)}$$

termination

$$Cl + ClCH_2CH_2 \xrightarrow{k_4} CH_2CHCl + HCl \qquad \text{Step (4)}$$
$$\text{(or other stable products)}$$

Step (1) requires a high activation energy, and its slowness would account for the existence of an induction period. Once a sufficiently high concentration of chlorine atoms is established however, decomposition of 1,2-dichloroethane would proceed mainly via the propagation steps which require a much lower activation energy. It is presumed that the propagation steps are, on the average, repeated many times before chlorine atoms are removed via the termination step.

A set of rate equations may be established involving the four steps in the proposed mechanism for the decomposition. In these equations:

M = concentration of $ClCH_2CH_2Cl$;

R_a = concentration of $ClCH_2CH_2$ radicals;

Cl = concentration of chlorine atoms;

R_b = concentration of $ClCH_2CHCl$ radicals.

Assuming reaction steps (1) and (3) to be of the first order, and steps (2) and (4) to be of the second order, the rate equations for the radical species in the system are as follows:

$$dR_a/dt = k_1 M - k_4 Cl R_a \tag{5–37}$$

$$dCl/dt = k_1 M - k_2 Cl M + k_3 R_b - k_4 Cl R_a \tag{5–38}$$

$$dR_b/dt = k_2 Cl M - k_3 R_b \tag{5–39}$$

The radical balance given by Eqn (5–40) also follows from the proposed mechanism:

$$R_a = Cl + R_b \tag{5–40}$$

After the induction period, it is assumed that approximately steady-state concentrations of the radical species are maintained while the decomposition proceeds. With this assumption, the differential terms on the left-hand side of Eqns (5–37) to (5–39) are equated to zero, and the three resultant equations plus Eqn (5–40) may be used to derive expressions for the concentrations of the radicals in terms of the rate coefficients, k_1 to k_4, and the concentration of 1,2-dichloroethane. These expressions are shown by Eqns (5–41) to (5–43).

$$R_a = [k_1 M (k_3 + k_2 M)/k_3 k_4]^{1/2} \tag{5–41}$$

$$Cl = [k_1 M/(k_4 + k_4 k_2 M/k_3)]^{1/2} \tag{5–42}$$

$$R_b = k_2 M [k_1 M/k_4 k_3 (k_3 + k_2 M)]^{1/2} \tag{5–43}$$

The rate of decomposition of 1,2-dichloroethane is given by Eqn (5–44) which may be expressed in the form of Eqn (5–45) by substituting for Cl (using Eqn (5–42)), and assuming that step (1) in the reaction scheme is very slow compared with step (2) (i.e. $k_1 M \ll k_2 Cl M$).

$$-dM/dt = k_1 M + k_2 Cl M \tag{5–44}$$

$$-dM/dt = M[(k_1 k_2 k_3)/k_4]^{1/2} [k_2 M/(k_3 + k_2 M)]^{1/2} \tag{5–45}$$

The last equation may be simplified to the form given by Eqn (5–46) if it is further assumed that the attack by chlorine atoms on the molecules of 1,2-dichloroethane is a much more facile process than the decomposition of the $ClCH_2CHCl$ radicals (i.e. $k_2 M \gg k_3$).

$$-dM/dt = M[(k_1 k_2 k_3)/k_4]^{1/2} \tag{5–46}$$

The mechanism proposed, with the associated assumptions, is therefore consistent with the decomposition of 1,2-dichloroethane being of the first order.

Estimates of the Arrhenius parameters for the individual steps (1) to (4), are available from independent studies on other reaction systems. These estimates, listed in Table 5–1 (K. D. King and E. S. Swinbourne, *Trans. Faraday Soc.*, 1970, **66**, 1145), may be used to provide a quantitative check on the validity of the proposed mechanism for the decomposition of 1,2-dichloroethane. Substitution of values from the table into the term, $[(k_1 k_2 k_3)/k_4]^{1/2}$ of Eqn (5–46), leads to the following estimate of the Arrhenius parameters for the decomposition: $E = 52\,000$ cal mole^{-1} and $\log A = 13\cdot3$. These estimates are in reasonable agreement with the experimental values reported by Goldfinger *et al.*, but differ somewhat from the values reported earlier by Barton and Howlett.

Table 5–1 Arrhenius parameters for radical-chain steps (pyrolysis of 1,2-dichloroethane)

The reaction steps are listed on p. 114. Units: E in cal mole^{-1}; A in s^{-1}, or s^{-1} litre mole^{-1}. Data are based on the values given by K. D. King and E. S. Swinbourne in *Trans. Faraday Soc.*, 1970, **66**, 1145.

Step	$\log A$	E
1	13·0	77 900
2	10·8	3 100
3	13·8	23 000
4	11·0	0
5	11·0	0
6	13·8	23 000

Checks of the type described, while of value in providing supporting evidence for a proposed radical-chain mechanism, must not be taken as conclusive, as more than one combination of reaction steps may sometimes lead to the required form of rate equation and to reasonable values for the Arrhenius parameters. For example, the alternative termination step involving chlorine atoms and $ClCH_2CHCl$ radicals (reaction step (5)) is also consistent with the first-order kinetics for the overall decomposition and leads to predicted values for the Arrhenius parameters similar to those reported by Goldfinger *et al.*

$$Cl + ClCH_2CHCl \rightarrow \textit{chain termination} \qquad\qquad \text{Step (5)}$$

A further feature to be considered in the overall scheme is that individual radicals of the type, $ClCH_2CH_2$, are as likely to dissociate via reaction step (6) as are the radicals, $ClCH_2CHCl$ via reaction step (3).

$$ClCH_2CH_2 \rightarrow Cl + CH_2CH_2 \qquad\qquad \text{Step (6)}$$

Allowance for this further possible reaction step would place the steady-state radical concentrations in the order, $Cl > R_b > R_a$, thus increasing the feasibility of step (5) being the dominant termination step. The reader may verify that a reaction sequence involving steps (1), (2), (3), (5), and (6) leads to kinetics of the first order for the decomposition of 1,2-dichloro-ethane, and to acceptable values for the Arrhenius parameters.

Other possible termination steps which might be considered in the reaction scheme for 1,2-dichloroethane are those involving pairs of chlorine atoms* or pairs of the larger radicals. With the unimolecular initiation step (1), these termination steps lead to rate expressions which are not of the first order and which have unacceptable values for the Arrhenius parameters. The steady-state hypothesis is therefore useful for the analysis of radical-chain reaction systems in that it simplifies the kinetic equations, and assists in the identification of those combinations of reaction steps which are meaningful in terms of the observed reaction behaviour. This process of identification is being continually improved with increasing knowledge of the characteristics of the individual reaction steps.

General classifications of steady-state radical-chain reaction systems in terms of various combinations of initiation and termination steps have been provided by P. Goldfinger, M. Letort, and M. Niclause (*Victor Henri Commemorative Volume, Contribution à l'Etude de la Structure Moleculaire*, 1948, p. 283. See also Friess *et al.*, pp. 458–9.[9]) Extensions of these classifications to more complex cases have been given by Stimson *et al.* (R. L. Failes and V. R. Stimson, *Austral. J. Chem.*, 1964, **17**, 851; R. L. Failes, V. R. Stimson, and E. S. Swinbourne, *Austral. J. Chem.*, 1965, **18**, 593).

5–5   Transient systems

This chapter has been concerned with methods of analysis of some typical steady-state systems which are of common occurrence in the field of chemical kinetics. Brief reference should, however, be made also to studies on transient and non-steady-state systems to which increasing attention has been given in recent years. These systems are of a specialized nature, and detailed discussion of them will not be attempted here; nevertheless, their analysis has yielded data of considerable value, particularly in the case of very fast reactions.

With conventional static systems, it is difficult to carry out accurate kinetic studies on a reaction which has a half-life of less than about 100 s. Flow methods are convenient for the study of reactions with half-lives ranging from 10^{-3} to 100 s, but, for the lower limit, a fast analytical method, such as spectrophotometry or electrical conducti-

---

* Unlikely to occur in the absence of a third body.

metry, is required. The accuracy and versatility of the flow method for studying rapid reactions have been considerably extended by the use of *accelerated-flow* and *stopped-flow* techniques, developed from the pioneering work of H. Hartridge, F. J. W. Roughton, Britton Chance, and others. A detailed discussion of these techniques is given in Friess *et al.*[9]

Transient reaction systems may be classified under two headings: those which occur naturally, and those which are artificially induced. Induction periods, degenerate chains, and spontaneous explosions constitute examples of naturally occurring transient behaviour. Transient situations may be artificially induced by rapidly perturbing an equilibrium or steady-state system; following the perturbation, studies may be made on the kinetics of the return of the system to its original state, or its traverse to a new equilibrium or steady-state condition. Such kinetics are called 'relaxation kinetics'. Techniques for introducing the perturbation include photolysis, a sudden change in temperature or pressure, ultrasonic absorption, and the use of high-frequency alternating electric fields. The methods of relaxation kinetics are particularly valuable for the study of fast reactions in solution. Thus, M. Eigen and his associates[17,18] by the use of such methods, have measured rate coefficients for proton transfer reactions with half-lives of the order of 10^{-10} to 10^{-11} mole^{-1} $1\,\text{s}^{-1}$.

As a simple example of relaxation kinetics, consider a system of the type,

$$A \underset{k_2}{\overset{k_1}{\rightleftharpoons}} B$$

which has been perturbed from its equilibrium state. If the opposing reactions are each of the first order, then the rate of return of species A to equilibrium conditions is given by Eqn (5–47) (see also Eqn (4–37) on page 85).

$$-dC_A/dt = k_1 C_A - k_2 C_B \tag{5–47}$$

If $x = C_A - C_{A\infty} = C_{B\infty} - C_B$, with $C_{A\infty}$ and $C_{B\infty}$ representing the equilibrium concentrations of A and B respectively, then Eqn (5–48) follows. 85).

$$-dx/dt = k_1(C_{A\infty} + x) - k_2(C_{B\infty} - x) \tag{5–48}$$

From the application of Eqn (5–47) under equilibrium conditions,

$$k_1 C_{A\infty} = k_2 C_{B\infty} \tag{5–49}$$

The simple form of Eqn (5–50) therefore follows.

$$-dx/dt = x(k_1 + k_2) \tag{5–50}$$

Assuming that at time, $t = 0$, $x = x_0$, integration yields Eqn (5–51).

$$\ln (x_0/x) = (k_1+k_2)t \tag{5-51}$$

If a relaxation time, t_r, is defined, such that $x_0/x_r = e$, substitution in Eqn (5–51) shows that the relaxation time is given by Eqn (5–52).

$$t_r = 1/(k_1 + k_2) \tag{5-52}$$

Thus, from the experimental determination of t_r for such a system, the sum of the rate coefficients, $k_1 + k_2$, may be estimated. If the equilibrium constant (equal to k_2/k_1) is also known, individual estimates of k_1 and k_2 may be calculated. It is interesting to note that, in this particular system, it is not necessary to measure absolute concentrations, as Eqn (5–51) is expressed in terms of the ratio, x_0/x, which is dimensionless.

Problem 5–6, at the end of this chapter, illustrates the application of relaxation methods to the chromate ion/dichromate ion system. Application of the method to other systems is described in the References.

Problems

5–1 The data given in Table 5–2 are for the thermal decomposition of gaseous di-t-butyl peroxide to acetone and ethane in a tubular flow-reactor of volume 82·4 ml at 188 °C. The decomposition follows first-order kinetics, and the stoichiometric equation is as follows:

$$(CH_3)_3CO_2 = 2(CH_3)_2CO + C_2H_6$$

A carrier gas has been used for the experiments, and any volume change resulting from the di-t-butyl peroxide decomposition may be assumed to be negligible. Using the form of Eqn (5–5), estimate the rate coefficient for the decomposition.

Table 5–2 Decomposition in a tubular flow-reactor

(See Problem 5–1)
The data are for the thermal decomposition of gaseous di-t-butyl peroxide. Symbols are as in Eqn (5–5). (Concentration in mole litre^{-1}; rate of flow in ml s^{-1}.)

u	$10^4 \times C_{A_0}$	$10^4 \times C_A$
2·31	8·17	6·35
1·20	9·25	5·41
1·00	8·71	4·67
0·88	5·90	2·82
0·66	10·17	4·00
0·56	6·58	2·24
0·51	6·00	1·80
0·44	4·81	1·21

5–2 A chemical substance, A, is decomposing in a tubular flow-reactor according to second-order kinetics, and there is no volume change resulting from the reaction.

(a) Under plug-flow conditions, the fraction decomposed after residence time, t, is given by Eqn (5–53)

$$(C_{A_0} - C_A)/C_{A_0} = C_{A_0}kt/(C_{A_0}kt + 1) \qquad (5\text{–}53)$$

Assuming $C_{A_0} = 1\cdot00$ mole l^{-1}, and a residence time of 100 s, estimate k if the fraction of A decomposed equals (i) $0\cdot25$, (ii) $0\cdot50$, or (iii) $0\cdot75$.

(b) Under laminar-flow conditions, with negligible diffusion, the mean fraction of A decomposed is given by Eqn (5–54) (Denbigh, p. 60).[1]

$$\frac{C_{A_0} - \overline{C}_A}{C_{A_0}} = C_{A_0}k\bar{t}\left[1 - \frac{C_{A_0}k\bar{t}}{2}\ln\left\{\frac{C_{A_0}k\bar{t} + 2}{C_{A_0}k\bar{t}}\right\}\right] \qquad (5\text{–}54)$$

In this equation, $\bar{t}$ represents the mean residence time, and $\overline{C}_A$ the mean concentration of A in the reagent material emerging from the reactor. As in part (a), estimate k if the mean fraction of A decomposed equals (i) $0\cdot25$, (ii) $0\cdot50$, or (iii) $0\cdot75$, assuming $C_{A_0} = 1\cdot00$ mole l^{-1} and $\bar{t} = 100$ s.

(c) From the values of k obtained in parts (a) and (b), estimate the ratio of the estimates of the rate coefficient assuming plug flow, to that assuming laminar flow, in each of the cases (i), (ii), and (iii) corresponding to the differing amounts of decomposition of A.

5–3   The hydrolysis of ethyl acetate by hydroxyl ion was studied in a stirred flow-reactor of volume 600 ml at 25 °C. Solutions of ethyl acetate ($0\cdot0401$ mole l^{-1}) and barium hydroxide (hydroxyl-ion concentration, $0\cdot00585$ mole l^{-1}) were introduced into the reactor at flow rates of $1\cdot19$ l h^{-1} and $1\cdot18$ l h^{-1} respectively, the concentration of hydroxyl ion at the outlet of the reactor being $0\cdot001092$ mole l^{-1}. Assuming that volume changes associated with the chemical process are negligible, and that the reaction is of the first order in each of ethyl acetate and hydroxyl ion, estimate the overall second-order rate coefficient for the reaction.

5–4   The decomposition of gaseous di-t-butyl peroxide in the presence of nitrogen carrier gas has been studied in a stirred flow-reactor by Mulcahy and Williams.[8] Table 5–3 lists data for the decomposition at temperatures near 481 °K.

(a) Using Eqn (5–36), estimate values for the rate coefficient at the recorded temperatures.

(b) Estimate values for k at 481 °K, assuming the simple Arrhenius relationship (Eqn (3–18)) with $E = 38\,300$ cal mole^{-1}.

(c) Express k at 481 °K in terms of a mean value and the standard deviation.

5–5   Gaseous neopentane thermally decomposes by a radical-chain process to produce predominantly methane and isobutene.

$$(CH_3)_4C = CH_2C(CH_3)_2 + CH_4$$

In the initial stages the order of the reaction is $1\cdot5$, and the Arrhenius parameters as measured at 500–570 °C are $E = 51\,500$ cal mole^{-1} and $A = 1\cdot5 \times 10^{13}$ mole$^{-1/2}$ ml$^{1/2}$ s^{-1} (J. Engel, A. Combe, M. Letort, and M. Niclause, *Compt. Rend.*, 1957, **244**, 453). It has been proposed that the decomposition proceeds by the following dominant steps:

initiation

$$(CH_3)_4C \xrightarrow{k_1} (CH_3)_3 + CH_3 \qquad \text{Step (1)}$$

propagation

$$CH_3 + (CH_3)_4C \xrightarrow{k_2} CH_4 + CH_2C(CH_3)_3 \qquad \text{Step (2)}$$

$$CH_2C(CH_3)_3 \xrightarrow{k_3} CH_2C(CH_3)_2 + CH_3 \qquad \text{Step (3)}$$

termination

$$2CH_3 \xrightarrow{k_4} C_2H_6 \qquad \text{Step (4)}$$

Table 5–3 Decomposition in a stirred flow-reactor

(See Problem 5–4)

The data are for the thermal decomposition of gaseous di-t-butyl peroxide.[8] $V_r = 276$ ml; P = total pressure (cm Hg) in reactor; T = temperature (°K); N_{A_0} = moles of reactant entering per second; N_c = moles of carrier gas entering per second; conv. = percentage of reactant converted to products. (Symbols are as in Eqn (5–36).)

T	P	$10^6 \times N_{A_0}$	N_c/N_{A_0}	$Conv.$
481·5	0·82	8·82	48·5	0·9
481·8	1·11	8·65	45·4	1·4
477·5	0·80	35·8	3·8	1·7
476·7	0·67	52·1	0	4·1
479·9	1·20	5·21	18·4	4·8
480·4	0·95	40·4	4·1	6·6
480·2	0·90	43·8	0	7·6
480·5	0·90	43·7	0	7·9
482·5	0·92	42·7	0	9·1
481·5	2·70	6·97	7·8	15·8
481·8	1·28	17·30	0	21·9
482·0	0·72	7·73	0	22·6
482·6	1·18	4·64	0	42·8
482·8	1·87	2·34	0	64·0

Table 5–4 Arrhenius parameters for radical-chain steps (pyrolysis of neopentane)

The reaction steps are as listed in Problem 5–5. Units: E in cal mole^{-1}; A in s^{-1}, or ml mole^{-1} s^{-1}.

$Step$	$\log A$	E
1	16·9	80 500
2	11·5	10 400
3	13·2	28 000
4	13·4	0

The Arrhenius parameters for these reaction steps are listed in Table 5–4.

(a) Show that the proposed mechanism is consistent with the overall decomposition being of order 1·5.

(b) Assuming the proposed mechanism, and using the data listed in Table 5–4, estimate the Arrhenius parameters for the overall decomposition process.

5–6 The application of relaxation methods to the chromate ion/dichromate ion system has been described by J. H. Swinehart.[19] The equilibria of interest are:

$$H^+ + CrO_4^{2-} \rightleftharpoons HCrO_4^- \qquad (1)$$

$$2HCrO_4^- \underset{k_r}{\overset{k_f}{\rightleftharpoons}} Cr_2O_7^{2-} + H_2O \qquad (2)$$

It may be assumed that (1) is a rapid equilibration compared with (2). Perturbation of the system may be accomplished by sudden dilution, following which the return towards equilibrium may be followed by measuring the change in pH with time. Under the experimental conditions described by Swinehart, the relaxation time is given by Eqn (5–55).

$$1/t_r = 4k_f C_{HCrO_4^-} + k_r C_{H_2O} \qquad (5\text{--}55)$$

The data listed in Table 5–5 were obtained from studies in dilute aqueous solutions at 23 °C. By means of a suitable plot, estimate k_f and k_r from the data provided.

Table 5–5   Relaxation studies on chromate/dichromate system
     (See Problem 5–6)

Experiment no.	$10^3 \times C_{HCrO_4^-}$ (mole litre^{-1})	Relaxation time (s)
1	0·60	29
2	0·55	26
3	2·15	26
4	1·40	25
5	3·10	25
6	3·95	20
7	5·73	17
8	6·20	16

5–7 A gas-phase reaction of the type,

$$A \to bB$$

is proceeding according to second-order kinetics at constant temperature and pressure in a static system. If V_0 is the original volume of reactant, and V the total volume of reaction mixture after time t, show that a straight-line plot should result when t is plotted against,

$$b(V_0 - V)/(V - bV_0) - 2\cdot303(1 - b)\log(V - bV_0)$$

5–8 The second-order rate coefficient, as defined by Eqn (5–56), for the gas-phase dimerization of butadiene to vinylcyclohexene at 326 °C equals $7\cdot1 \times 10^{-3}$ mole^{-1} l s^{-1} (see page 79).

$$-dC_B/dt = 2kC_B^2 \qquad (5\text{--}56)$$

(C_B is the concentration of butadiene.)
    The dimerization reaction is to be conducted in a tubular flow-reactor operating under plug flow conditions, with an initial rate of flow of 0·01 l s^{-1} for the pure reagent entering the reactor at one atmosphere pressure. The pressure may be assumed uniform within the reactor.
(a) What volume of reaction vessel would be required for the dimerization to proceed to 50% completion?
(b) If the process were not accompanied by a volume change, what would then be the volume of reactor required for 50% reaction?

References

References to flow systems

1. Denbigh, K. *Chemical Reactor Theory. An Introduction.* Cambridge University Press, London, 1965.
2. Hougen, O. A., and K. M. Watson. *Chemical Process Principles. Part 3: Kinetics and Catalysis.* John Wiley, New York, 1947.
3. Aris, R. *Introduction to the Analysis of Chemical Reactors.* Prentice-Hall, Englewood Cliffs, N.J., 1965.
4. Penner, S. S. *Chemical Reactions in Flow Systems.* Butterworths, London, 1955.
5. Herndon, W. C. Kinetics in gas-phase stirred-flow reactors. *J. Chem. Ed.,* 1964, **41**, 425.
6. Denbigh, K. G. *Trans. Faraday Soc.,* 1944, **40**, 352, and 1947, **43**, 648; B. Stead, F. M. Page, and K. G. Denbigh, *Disc. Faraday Soc.,* 1947, **2**, 263; K. G. Denbigh, M. Hicks, and F. M. Page, *Trans. Faraday Soc.,* 1948, **44**, 479. A summarizing paper on stirred-flow systems has been given by K. G. Denbigh and F. M. Page in *Trans. Faraday Soc.,* 1954, **50**, 145.
7. Taylor, J. E. The stirred flow technique. Mathematically developed applications for complex kinetic reactions, *J. Chem. Ed.,* 1969, **46**, 742.
8. Mulcahy, M. F. R., and D. J. Williams. A stirred-flow reactor for investigating the kinetics of gaseous reactions. *Austral. J. Chem.,* 1961, **14**, 534.
9. Friess, S. L., E. S. Lewis, and A. Weissberger (Eds.). *Investigation of Rates and Mechanisms of Reactions* (vol. 8 of *Technique of Organic Chemistry*). John Wiley (Interscience), New York, 2nd ed., 1961. See particularly Chapter 14 (by F. J. W. Roughton and Britton Chance) in Part 2.

References to chain reactions

10. Dainton, F. S. *Chain Reactions. An Introduction.* Methuen, London, 2nd ed., 1966.
11. Frost, A. A., and R. G. Pearson. *Kinetics and Mechanism.* John Wiley, New York, 2nd ed., 1961. Chapter 10.
12. Trotman-Dickenson, A. F. *Free Radicals.* Methuen, London, 1959.
13. Steacie, E. W. R. *Atomic and Free Radical Reactions.* Reinhold, New York, 2nd ed., 1954.
14. Semenoff, N. *Chemical Kinetics and Chain Reactions.* Clarendon Press, Oxford, 1935.
15. See also the series, *Advances in Free Radicals,* Academic Press, New York, vol. 1 (1967) and subsequent volumes.

References to relaxation kinetics and related topics

16. Caldin, E. F. *Fast Reactions in Solution.* John Wiley, New York, 1964.
17. Eigen, M., and L. De Maeyer. Chapter 18 in *Investigation of Rates and Mechanisms of Reactions* (see reference 9, above). This volume gives a comprehensive coverage by a number of other contributors to the study of very rapid reactions.
18. Eigen, M. *Disc. Faraday Soc.,* 1954, **17**, 194, and subsequent papers. See also M. Eigen and J. S. Johnson, *Ann. Rev. Phys. Chem.,* 1960, **11**, 307.
19. Swinehart, J. H. Relaxation kinetics. *J. Chem. Ed.,* 1967, **44**, 524.
20. Finholt, J. E. The temperature-jump method for the study of fast reactions. *J. Chem. Ed.,* 1968, **45**, 394.

Index